Glencoe

CHEMISTRY

MATTER AND CHANGE

Solving Problems: A Chemistry Handbook

Glencoe McGraw-Hill

New York, New York   Columbus, Ohio   Woodland Hills, California   Peoria, Illinois

Glencoe

CHEMISTRY
MATTER AND CHANGE

Hands-On Learning:
Laboratory Manual, SE/TE
Forensics Laboratory Manual, SE/TE
CBL Laboratory Manual, SE/TE
Small-Scale Laboratory Manual,
  SE/TE
ChemLab and MiniLab Worksheets

Review/Reinforcement:
Study Guide for Content Mastery,
  SE/TE
Solving Problems: A Chemistry
  Handbook
Reviewing Chemistry
Guided Reading Audio Program

Applications and Enrichment:
Challenge Problems
Supplemental Problems

Assessment:
Chapter Assessment
MindJogger Videoquizzes
  (VHS/DVD)
TestCheck Software,
  Windows/MacIntosh

Teacher Resources:
Lesson Plans
Block Scheduling Lesson Plans
Spanish Resources
Section Focus Transparencies and
  Masters
Math Skills Transparencies and
  Masters
Teaching Transparencies and Masters
Solutions Manual

Technology:
Chemistry Interactive CD-ROM
Vocabulary PuzzleMaker Software,
  Windows/MacIntosh
Glencoe Science Web site:
  science.glencoe.com

Send all inquiries to:
Glencoe/McGraw-Hill
8787 Orion Place
Columbus, OH 43240-4027

ISBN 0-07-824536-2
Printed in the United States of America.
  10 045 09 08 07 06 05

To the Teacher

Solving Problems: A Chemistry Handbook provides not only practice but guidance in how to solve problems in chemistry. This handbook covers the main concepts in each section of **Chemistry: Matter and Change**. The text material is brief; the chapters focus instead on the example problems, practice problems, and other questions that reinforce students' knowledge and problem-solving skills. Answers to the problems and questions are found at the back of the book. *Solving Problems: A Chemistry Handbook* is a powerful tool for independent study, reteaching, and review.

Contents

Introduction to Chemistry

1.1 The Stories of Two Chemicals

A chemical is any substance that has a definite composition. Ozone is a chemical that is made up of three particles of oxygen. Ozone forms a thick blanket above the clouds in the stratosphere. This layer of ozone protects Earth from overexposure to ultraviolet radiation from the Sun. You are probably familiar with the damage that exposure to ultraviolet radiation can do to your skin in the form of sunburn. Ultraviolet radiation can also harm other animals and plants. In the 1980s, scientists documented that the ozone layer around Earth was becoming measurably thinner in some spots.

In the 1970s, scientists had observed that large quantities of chlorofluorocarbons (CFCs) had accumulated in Earth's atmosphere. CFCs are chemicals that contain chlorine, fluorine, and carbon. CFCs were used as coolants in refrigerators and air conditioners and as propellants in spray cans because they were considered relatively nonreactive. Some scientists hypothesized that there might be a connection between the concentration of CFCs in the atmosphere and the thinning of the ozone layer.

1.2 Chemistry and Matter

Chemistry is the study of matter and the changes that it undergoes. **Matter** is anything that has mass and takes up space. **Mass** is a measurement of the amount of matter in an object. Everything, however, is not made of matter. For example, heat, light, radio waves, and magnetic fields are some things that are not made of matter.

You might wonder why scientists measure matter in terms of mass, and not in terms of weight. Your body is made of matter, and you probably weigh yourself in pounds. However, your **weight** is not just a measure of the amount of matter in your body. Your weight also includes the effect of Earth's gravitational pull on your body. This force is not the same everywhere on Earth. Scientists use mass to measure matter instead of weight because they need to compare measurements taken in different locations.

Matter is made up of particles, called atoms, that are so small they cannot be seen with an ordinary light microscope. The structure, composition, and behavior of all matter can be explained by atoms and the changes they undergo.

Because there are so many types of matter, there are many areas of study in the field of chemistry. Chemistry is usually divided into five branches, as summarized in the table below.

Branches of Chemistry		
Branch	**Area of emphasis**	**Examples**
Organic chemistry	most carbon-containing chemicals	pharmaceuticals, plastics
Inorganic chemistry	in general, matter that does not contain carbon	minerals, metals and nonmetals, semiconductors
Physical chemistry	the behavior and changes of matter and the related energy changes	reaction rates, reaction mechanisms
Analytical chemistry	components and composition of substances	food nutrients, quality control
Biochemistry	matter and processes of living organisms	metabolism, fermentation

1.3 Scientific Methods

A **scientific method** is a systematic approach used to answer a question or study a situation. It is both an organized way for scientists to do research and a way for scientists to verify the work of other scientists. A typical scientific method includes making observations, forming a hypothesis, performing an experiment, and arriving at a conclusion.

Scientific study usually begins with observations. Often, a scientist will begin with **qualitative data**—information that describes color, odor, shape, or some other physical characteristic that relates to the five senses. Chemists also use **quantitative data.** This type of data is numerical. It tells how much, how little, how big, or how fast.

Practice Problems

1. Identify each of the following as an example of qualitative data or quantitative data.

   a. taste of an apple
   b. mass of a brick
   c. speed of a car

   d. length of a rod
   e. texture of a leaf
   f. weight of an elephant

A **hypothesis** is a possible explanation for what has been observed. Based on the observations of ozone thinning and CFC buildup in the atmosphere, the chemists Mario Molina and F. Sherwood Rowland hypothesized that CFCs break down in the atmosphere due to the Sun's ultraviolet rays. They further hypothesized that a chlorine particle produced by the breakdown of CFCs could break down ozone.

An **experiment** is a set of controlled observations that test a hypothesis. In an experiment, a scientist will set up and change one variable at a time. A variable is a quantity that can have more than one value. The variable that is changed in an experiment is called the **independent variable.** The variable that you watch to see how it changes as a result of your changes to the independent variable is called the **dependent variable.** For example, if you wanted to test the effect of fertilizer on plant growth, you would change the amount of fertilizer applied to the same kinds of plants. The amount of fertilizer applied would be the independent variable in this experiment. Plant growth would be the dependent variable. Many experiments also include a **control,** which is a standard for comparison; in this case, plants to which no fertilizer is applied.

A **conclusion** is a judgment based on the data obtained in the experiment. If data support a hypothesis, the hypothesis is tentatively affirmed. Hypotheses are never proven; they are always subject to additional research. If additional data do not support a hypothesis, the hypothesis is discarded or modified. Most hypotheses are not supported by data. Whether the hypothesis is supported or not, the data collected may still be useful. Over time, data from many experiments can be used to form a visual, verbal, and/or mathematical explanation—called a **model**—of the phenomenon being studied.

A **theory** is an explanation that has been supported by many experiments. Theories state broad principles of nature. Although theories are the best explanations of phenomena that scientists have at any given time, they are always subject to new experimental data and are modified to include new data.

A **scientific law** describes a relationship in nature that is supported by many experiments and for which no exception has been found. Scientists may use models and theories to explain why this relationship exists.

1.4 Scientific Research

Pure research is done to gain knowledge for the sake of knowledge itself. Molina and Rowland's research on the behavior of CFCs—showing that in the lab CFCs could speed up the breakdown of ozone—was motivated by their curiosity and is an example of pure research. **Applied research** is undertaken to solve a specific problem. Scientists are conducting experiments to find chemicals to replace CFCs. These experiments are examples of applied research.

▶ **Laboratory safety** During your study of chemistry, you will conduct experiments in the laboratory. When working in the lab, you are responsible for the safety of yourself and others working around you. Each time you enter the lab, use these safety rules as a guide.

Safety in the Laboratory
1. Study your lab assignment before you come to the lab. If you have any questions, be sure to ask your teacher for help.
2. Do not perform experiments without your teacher's permission. **Never** work alone in the laboratory.
3. Use the table on the inside front cover of this textbook to understand the safety symbols. Read all **CAUTION** statements.
4. Safety goggles and a laboratory apron must be worn whenever you are in the lab. Gloves should be worn whenever you use chemicals that cause irritations or can be absorbed through the skin. Long hair must be tied back.
5. Do not wear contact lenses in the lab, even under goggles. Lenses can absorb vapors and are difficult to remove in case of an emergency.
6. Avoid wearing loose, draping clothing and dangling jewelry. Bare feet and sandals are not permitted in the lab.
7. Eating, drinking, and chewing gum are not allowed in the lab.
8. Know where to find and how to use the fire extinguisher, safety shower, fire blanket, and first-aid kit.

Safety in the Laboratory, *continued*

9. Report any accident, injury, incorrect procedure, or damaged equipment to your teacher.

10. If chemicals come in contact with your eyes or skin, flush the area immediately with large quantities of water. Immediately inform your teacher of the nature of the spill.

11. Handle all chemicals carefully. Check the labels of all bottles **before** removing the contents. Read the label three times:
    - Before you pick up the container.
    - When the container is in your hand.
    - When you put the bottle back.

12. Do not take reagent bottles to your work area unless instructed to do so. Use test tubes, paper, or beakers to obtain your chemicals. Take only small amounts. It is easier to get more than to dispose of excess.

13. Do not return unused chemicals to the stock bottle.

14. Do not insert droppers into reagent bottles. Pour a small amount of the chemical into a beaker.

15. **Never** taste any chemicals. **Never** draw any chemicals into a pipette with your mouth.

16. Keep combustible materials away from open flames.

17. Handle toxic and combustible gases only under the direction of your teacher. Use the fume hood when such materials are present.

18. When heating a substance in a test tube, be careful not to point the mouth of the test tube at another person or yourself. Never look down the mouth of a test tube.

19. Do not heat graduated cylinders, burettes, or pipettes with a laboratory burner.

20. Use caution and proper equipment when handling hot apparatus or glassware. Hot glass looks the same as cool glass.

21. Dispose of broken glass, unused chemicals, and products of reactions only as directed by your teacher.

22. Know the correct procedure for preparing acid solutions. **Always** add the acid slowly to the water.

23. Keep the balance area clean. Never place chemicals directly on the pan of a balance.

24. After completing an experiment, clean and put away your equipment. Clean your work area. Make sure the gas and water are turned off. Wash your hands with soap and water before you leave the lab.

Chapter 1 Review

2. How does the ozone layer protect Earth?

3. Why did scientists think that the thinning of the ozone layer might be related to CFCs?

4. Contrast mass and weight.

5. During a chemistry lab, a student noted the following data about an unknown chemical she was studying: colorless, dissolves in water at room temperature, melts at 95°C, boils at 800°C. Classify each piece of data as either qualitative data or quantitative data.

6. Identify the dependent variable and the independent variable in the following experiments.

  a. A student tests the ability of a given chemical to dissolve in water at three different temperatures.

  b. A farmer compares how his crops grow with and without phosphorous fertilizers.

  c. An environmentalist tests the acidity of water samples at five different distances from a factory.

7. Explain why hypotheses and theories are always tentative explanations.

8. List two possible hypotheses about the relationship between ozone and CFCs.

9. Classify each kind of research as either pure or applied.

  a. A scientist studies plants in a rain forest in search of chemicals that might be used to treat AIDS.

  b. A researcher studies the effects of hormones on the brain of a worm.

  c. A researcher tries to develop cleaner burning fuels to help reduce air pollution.

10. State two rules you should follow when handling chemicals.

11. How should you dispose of the following items in the lab: broken glass, products of chemical reactions, unused chemicals?

Data Analysis

2.1 Units of Measurement

You probably know your height in feet and inches. Most people outside the United States, however, measure height in meters and centimeters. The system of standard units that includes the meter is called the metric system. Scientists today use a revised form of the metric system called the Système Internationale d'Unités, or SI.

▶ **Base units** There are seven base units in SI. A **base unit** is a unit of measure that is based on an object or event in the physical world. Table 2-1 lists the seven SI base units, their abbreviations, and the quantities they are used to measure.

Table 2-1

SI Base Units	
Quantity	**Base unit**
Time	second (s)
Length	meter (m)
Mass	kilogram (kg)
Temperature	kelvin (K)
Amount of a substance	mole (mol)
Electric current	ampere (A)
Luminous intensity	candela (cd)

SI is based on a decimal system. So are the prefixes in Table 2-2, which are used to extend the range of SI units.

Table 2-2

Prefixes Used with SI Units				
Prefix	Symbol	Factor	Scientific notation	Example
giga	G	1 000 000 000	10^9	gigameter (Gm)
mega	M	1 000 000	10^6	megagram (Mg)
kilo	k	1000	10^3	kilometer (km)
deci	d	1/10	10^{-1}	deciliter (dL)
centi	c	1/100	10^{-2}	centimeter (cm)
milli	m	1/1000	10^{-3}	milligram (mg)
micro	μ	1/1 000 000	10^{-6}	microgram (μg)
nano	n	1/1 000 000 000	10^{-9}	nanometer (nm)
pico	p	1/1 000 000 000 000	10^{-12}	picometer (pm)

Example Problem 2-1
Using Prefixes with SI Units

How many picograms are in a gram?

The prefix *pico-* means 10^{-12}, or 1/1 000 000 000 000. Thus, there are 10^{12}, or 1 000 000 000 000, picograms in one gram.

Practice Problems
1. How many centigrams are in a gram?
2. How many liters are in a kiloliter?
3. How many nanoseconds are in a second?
4. How many meters are in a kilometer?

▶ **Derived units** Not all quantities can be measured using SI base units. For example, volume and density are measured using units that are a combination of base units. An SI unit that is defined by a combination of base units is called a **derived unit.** The SI unit for volume is the liter. A **liter** is a cubic meter, that is, a cube whose sides are all one meter in length. **Density** is a ratio that compares the mass of an object to its volume. The SI units for density are often grams per cubic centimeter (g/cm^3) or grams per milliliter (g/mL). One centimeter cubed is equivalent to one milliliter.

Example Problem 2-2
Calculating Density

A 1.1-g ice cube raises the level of water in a 10-mL graduated cylinder 1.2 mL. What is the density of the ice cube?

To find the ice cube's density, divide its mass by the volume of water it displaced and solve.

density = mass/volume

$$density = \frac{1.1 \text{ g}}{1.2 \text{ mL}} = 0.92 \text{ g/mL}$$

Example Problem 2-3
Using Density and Volume to Find Mass

Suppose you drop a solid gold cube into a 10-mL graduated cylinder containing 8.50 mL of water. The level of the water rises to 10.70 mL. You know that gold has a density of 19.3 g/cm^3, or 19.3 g/mL. What is the mass of the gold cube?

To find the mass of the gold cube, rearrange the equation for density to solve for mass.

density = mass/volume

mass = volume × density

Substitute the values for volume and density into the equation and solve for mass.

mass = 2.20 mL × 19.3 g/mL = 42.5 g

Practice Problems

5. Calculate the density of a piece of bone with a mass of 3.8 g and a volume of 2.0 cm^3.

6. A spoonful of sugar with a mass of 8.8 grams is poured into a 10-mL graduated cylinder. The volume reading is 5.5 mL. What is the density of the sugar?

7. A 10.0-gram pat of butter raises the water level in a 50-mL graduated cylinder by 11.6 mL. What is the density of the butter?

8. A sample of metal has a mass of 34.65 g. When placed in a graduated cylinder containing water, the water level rises 3.3 mL. Which of the following metals is the sample made from: silver, which has a density of 10.5 g/cm^3; tin, which has a density of 7.28 g/cm^3; or titanium, which has a density of 4.5 g/cm^3?

9. Rock salt has a density of 2.18 g/cm^3. What would the volume be of a 4.8-g sample of rock salt?

10. A piece of lead displaces 1.5 mL of water in a graduated cylinder. Lead has a density of 11.34 g/cm^3. What is the mass of the piece of lead?

▶ **Temperature** The temperature of an object describes how hot or cold the object is relative to other objects. Scientists use two temperature scales—the Celsius scale and the Kelvin scale—to measure temperature. You will be using the Celsius scale in most of your experiments. On the Celsius scale, the freezing point of water is defined as 0 degrees and the boiling point of water is defined as 100 degrees.

A **kelvin** is the SI base unit of temperature. On the Kelvin scale, water freezes at about 273 K and boils at about 373 K. One kelvin is equal in size to one degree on the Celsius scale. To convert from degrees Celsius to kelvins, add 273 to the Celsius measurement. To convert from kelvins to degrees Celsius, subtract 273 from the measurement in kelvins.

Practice Problems

11. Convert each temperature reported in degrees Celsius to kelvins.
    a. 54°C
    b. −54°C
    c. 15°C

12. Convert each temperature reported in kelvins to degrees Celsius.
    a. 32 K
    b. 0 K
    c. 281 K

2.2 Scientific Notation and Dimensional Analysis

Extremely small and extremely large numbers can be compared more easily when they are converted into a form called scientific notation. **Scientific notation** expresses numbers as a multiple of two factors: a number between 1 and 10; and ten raised to a power, or exponent. The exponent tells you how many times the first factor must be multiplied by ten. When numbers larger than 1 are expressed in scientific notation, the power of ten is positive. When numbers smaller than 1 are expressed in scientific notation, the power of ten is negative. For example, 2000 is written as 2×10^3 in scientific notation, and 0.002 is written as 2×10^{-3}.

Example Problem 2-4
Expressing Quantities in Scientific Notation

The surface area of the Pacific Ocean is 166 000 000 000 000 m^2. Write this quantity in scientific notation.

To write the quantity in scientific notation, move the decimal point to after the first digit to produce a factor that is between 1 and 10. Then count the number of places you moved the decimal point; this number is the exponent (n). Delete the extra zeros at the end of the first factor, and multiply the result by 10^n. When the decimal point moves to the left, n is positive. When the decimal point moves to the right, n is negative. In this problem, the decimal point moves 14 places to the left; thus, the quantity is written as 1.66×10^{14} in scientific notation.

Practice Problems

13. Express the following quantities in scientific notation.

   a. 50 000 m/s^2

   b. 0.000 000 000 62 kg

   c. 0.000 023 s

   d. 21 300 000 mL

   e. 990 900 000 m/s

   f. 0.000 000 004 L

▶ **Adding and subtracting using scientific notation** To add or subtract quantities written in scientific notation, the quantities must have the same exponent. For example, 4.5×10^{14} m + 2.1×10^{14} m = 6.6×10^{14} m. If two quantities are expressed to different powers of ten, you must change one of the quantities so that they are both expressed to the same power of ten before you add or subtract them.

Example Problem 2-5
Adding Quantities Written in Scientific Notation

Solve the following problem.

2.45×10^{14} kg + 4.00×10^{12} kg

First express both quantities to the same power of ten. Either quantity can be changed. For example, you might change 2.45×10^{14} to 245×10^{12}. Then add the quantities: 245×10^{12} kg + 4.00×10^{12} kg = 249×10^{12} kg. Write the final answer in scientific notation: 2.49×10^{14} kg.

Practice Problems
14. Solve the following addition and subtraction problems. Write your answers in scientific notation.
   a. $5.10 \times 10^{20} + 4.11 \times 10^{21}$
   b. $6.20 \times 10^{8} - 3.0 \times 10^{6}$
   c. $2.303 \times 10^{5} - 2.30 \times 10^{3}$
   d. $1.20 \times 10^{-4} + 4.7 \times 10^{-5}$
   e. $6.20 \times 10^{-6} + 5.30 \times 10^{-5}$
   f. $8.200 \times 10^{2} - 2.0 \times 10^{-1}$

▶ **Multiplying and dividing using scientific notation** When multiplying or dividing quantities written in scientific notation, the quantities do not have to have the same exponent. For multiplication, multiply the first factors, then add the exponents. For division, divide the first factors, then subtract the exponents.

Example Problem 2-6
Multiplying Quantities Written in Scientific Notation

Solve the following problem.

$(2 \times 10^{14}$ cm$) \times (4 \times 10^{12}$ cm$)$

To solve the multiplication problem, first multiply the factors: $2 \times 4 = 8$. Then add the exponents: $14 + 12 = 26$. Combine the factors: 8×10^{26}. Finally, multiply the units and write your answer in scientific notation: 8×10^{26} cm^2.

Practice Problems

15. Solve the following multiplication and division problems. Write your answers in scientific notation.

   a. $(12 \times 10^4 \text{ m}) \times (5 \times 10^{-2} \text{ m})$

   b. $(3 \times 10^7 \text{ km}) \times (3 \times 10^7 \text{ km})$

   c. $(2 \times 10^{-4} \text{ mm}) \times (2 \times 10^{-4} \text{ mm})$

   d. $(90 \times 10^{14} \text{ kg}) \div (9 \times 10^{12} \text{ L})$

   e. $(12 \times 10^{-4} \text{ m}) \div (3 \times 10^{-4} \text{ s})$

   f. $(20 \times 10^{15} \text{ km}) \div (5 \times 10^{11} \text{ s})$

▶ **Dimensional analysis** Dimensional analysis is a method of problem solving that focuses on the units that are used to describe matter. Dimensional analysis often uses conversion factors. A **conversion factor** is a ratio of equivalent values used to express the same quantity in different units. A conversion factor is always equal to 1. Multiplying a quantity by a conversion factor does not change its value—because it is the same as multiplying by 1—but the units of the quantity can change.

Example Problem 2-7
Converting From One Unit to Another Unit

How many centigrams are in 5 kilograms?

Two conversion factors are needed to solve this problem. Remember that there are 1000 grams in a kilogram and 100 centigrams in a gram. To determine the number of centigrams in 1 kilogram, set up the first conversion factor so that kilograms cancel out. Set up the second conversion factor so that grams cancel out.

$$5 \text{ kg} \times \frac{1 \text{ g}}{1000 \text{ kg}} \times \frac{100 \text{ cg}}{1 \text{ g}} = 0.5 \text{ cg}$$

Practice Problems

16. Mount Everest is 8847 m high. How many centimeters high is the mountain?

17. Your friend is 1.56 m tall. How many millimeters tall is your friend?

18. A family consumes 2.5 gallons of milk per week. How many liters of milk do they need to buy for one week?
(Hint: 1 L = 0.908 quart; 1 gallon = 4 quarts.)

19. How many hours are there in one week? How many minutes are there in one week?

2.3 How reliable are measurements?

When scientists look at measurements, they want to know how accurate as well as how precise the measurements are. **Accuracy** refers to how close a measured value is to an accepted value. **Precision** refers to how close a series of measurements are to one another. Precise measurements might not be accurate, and accurate measurements might not be precise. When you make measurements, you want to aim for both precision and accuracy.

▶ **Percent error** Quantities measured during an experiment are called experimental values. The difference between an accepted value and an experimental value is called an error. The ratio of an error to an accepted value is called **percent error.** The equation for percent error is as follows.

$$\text{Percent error} = \frac{\text{error}}{\text{accepted value}} \times 100$$

When you calculate percent error, ignore any plus or minus signs because only the size of the error counts.

Example Problem 2-8
Calculating Percent Error

Juan calculated the density of aluminum three times.

Trial 1: 2.74 g/cm^3

Trial 2: 2.68 g/cm^3

Trial 3: 2.84 g/cm^3

Aluminum has a density of 2.70 g/cm³. Calculate the percent error for each trial.

First, calculate the error for each trial by subtracting Juan's measurement from the accepted value (2.70 g/cm³).

Trial 1: error = 2.70 g/cm³ − 2.74 g/cm³ = −0.04 g/cm³

Trial 2: error = 2.70 g/cm³ − 2.68 g/cm³ = 0.02 g/cm³

Trial 3: error = 2.70 g/cm³ − 2.84 g/cm³ = −0.14 g/cm³

Then, substitute each error and the accepted value into the percent error equation. Ignore the plus and minus signs.

Trial 1: percent error = $\dfrac{0.04 \text{ g/cm}^3}{2.70 \text{ g/cm}^3} \times 100 = 1.48\%$

Trial 2: percent error = $\dfrac{0.02 \text{ g/cm}^3}{2.70 \text{ g/cm}^3} \times 100 = 0.741\%$

Trial 3: percent error = $\dfrac{0.14 \text{ g/cm}^3}{2.70 \text{ g/cm}^3} \times 100 = 5.19\%$

Practice Problems

20. Suppose you calculate your semester grade in chemistry as 90.1, but you receive a grade of 89.4. What is your percent error?

21. On a bathroom scale, a person always weighs 2.5 pounds less than on the scale at the doctor's office. What is the percent error of the bathroom scale if the person's actual weight is 125 pounds?

22. A length of wood has a labeled length value of 2.50 meters. You measure its length three times. Each time you get the same value: 2.35 meters.

a. What is the percent error of your measurements?

b. Are your measurements precise? Are they accurate?

▶ **Significant figures** The number of digits reported in a measurement indicates how precise the measurement is. The more digits reported, the more precise the measurement. The digits reported in a measurement are called significant figures. **Significant figures** include all known digits plus one estimated digit.

These rules will help you recognize significant figures.

1. Nonzero numbers are always significant.

   45.893421 min has eight significant figures

2. Zeros between nonzero numbers are always significant.

   2001.5 km has five significant figures

3. All final zeros to the right of the decimal place are significant.

   6.00 g has three significant figures

4. Zeros that act as placeholders are not significant. You can convert quantities to scientific notation to remove placeholder zeros.

   0.0089 g and 290 g each have two significant figures

5. Counting numbers and defined constants have an infinite number of significant figures.

Example Problem 2-9
Counting Significant Figures

How many significant figures are in the following measurements?

   a. 0.002 849 kg
   b. 40 030 kg

Apply rules 1–4 from above. Check your answers by writing the quantities in scientific notation.

   a. 0.002 849 kg has four significant figures; 2.849×10^{-3}
   b. 40 030 kg has four significant figures; 4.003×10^4

Practice Problems

23. Determine the number of significant figures in each measurement.

   a. 0.000 010 L   **c.** 2.4050×10^{-4} kg
   b. 907.0 km   **d.** 300 100 000 g

▶ **Rounding off numbers** When you report a calculation, your answer should have no more significant figures than the piece of data you used in your calculation with the fewest number of significant figures. Thus, if you calculate the density of an object with a mass of 12.33 g and a volume of 19.1 cm^3, your answer should have only three significant figures. However, when you divide these quantities using your calculator, it will display 0.6455497—many more figures than you can report in your answer. You will have to round off the number to three significant figures, or 0.646.

Here are some rules to help you round off numbers.

1. If the digit to the immediate right of the last significant figure is less than five, do not change the last significant figure.

2. If the digit to the immediate right of the last significant figure is greater than five, round up the last significant figure.

3. If the digit to the immediate right of the last significant figure is equal to five and is followed by a nonzero digit, round up the last significant figure.

4. If the digit to the immediate right of the last significant figure is equal to five and is not followed by a nonzero digit, look at the last significant figure. If it is an odd digit, round it up. If it is an even digit, do not round up.

Whether you are adding, subtracting, multiplying, or dividing, you must always report your answer so that it has the same number of significant figures as the measurement with the fewest significant figures.

Example Problem 2-10
Rounding Off Numbers

Round the following number to three significant figures: 3.4650.

Rule 4 applies. The digit to the immediate right of the last significant figure is a 5 followed by a zero. Because the last significant figure is an even digit (6), do not round up. The answer is 3.46.

Practice Problems

24. Round each number to five significant figures. Write your answers in scientific notation.

  a. 0.000 249 950

  b. 907.0759

  c. 24 501 759

  d. 300 100 500

25. Complete the following calculations. Round off your answers as needed.

  a. 52.6 g + 309.1 g + 77.214 g

  b. 927.37 mL − 231.458 mL

  c. 245.01 km × 2.1 km

  d. 529.31 m ÷ 0.9000 s

2.4 Representing Data

A **graph** is a visual display of data. Representing your data in graphs can reveal a pattern if one exists. You will encounter several different kinds of graphs in your study of chemistry.

▶ **Circle graphs** A circle graph is used to show the parts of a fixed whole. This kind of graph is sometimes called a pie chart because it is a circle divided into wedges that look like pieces of pie. Each wedge represents a percentage of the whole. The entire graph represents 100 percent.

▶ **Bar graphs** A bar graph is often used to show how a quantity varies with time, location, or temperature. In this situation, the quantity being measured appears on the vertical axis. The independent variable—time, for example—appears on the horizontal axis.

▶ **Line graphs** The points on a line graph represent the intersection of data for two variables. The independent variable is plotted on the horizontal axis. The dependent variable is plotted on the vertical axis. The points on a line graph are connected by a best fit line, which is a line drawn so that as many points fall above the line as below it.

If a best fit line is straight, there is a linear relationship between the variables. This relationship can be described by the steepness, or slope, of the line. If the line rises to the right, the slope is positive. A positive slope indicates that the dependent variable increases as the independent variable increases. If the line falls to the right, the slope is negative. A negative slope indicates that the dependent variable decreases as the independent variable increases. You can use two data points to calculate the slope of a line.

Example Problem 2-11
Calculating the Slope of a Line from Data Points

Calculate the slope of a line that contains these data points: (3.0 cm^3, 6.0 g) and (12 cm^3, 24 g).

To calculate the slope of a line from data points, substitute the values into the following equation.

$$\text{slope} = \frac{y_2 - y_1}{x_2 - x_1} = \Delta y/\Delta x$$

$$\text{slope} = \frac{24 \text{ g} - 6.0 \text{ g}}{12 \text{ cm}^3 - 3.0 \text{ cm}^3} = \frac{18 \text{ g}}{9.0 \text{ cm}^3} = 2.0 \text{ g/cm}^3$$

Practice Problems
26. Calculate the slope of each line using the points given.
 a. (24 cm^3, 36 g), (12 cm^3, 18 g)
 b. (25.6 cm^3, 28.16 g), (17.3 cm^3, 19.03 g)
 c. (15s, 147 m), (21 s, 205.8 m)
 d. (55 kJ, $18.75°$C), (75 kJ, $75.00°$C)

▶ **Interpreting data** When you are asked to read the information from a graph, first identify the dependent variable and the independent variable. Look at the ranges of the data and think about what measurements were taken to obtain the data. Determine whether the relationship between the variables is linear or nonlinear. If the relationship is linear, determine if the slope is positive or negative.

If the points on the graph are connected, they are considered continuous. You can read data that falls between the measured points. This process is called interpolation. You can extend the line on a graph beyond the plotted points and estimate values for the variables. This process is called extrapolation. Extrapolation is less reliable than interpolation because you are going beyond the range of the data collected.

Chapter 2 Review

27. Which SI units would you use to measure the following quantities?

    a. the amount of water you drink in one day

    b. the distance from New York to San Francisco

    c. the mass of an apple

28. How does adding the prefix *kilo-* to an SI unit affect the quantity being described?

29. What units are used for density in the SI system? Are these base units or derived units? Explain your answer.

30. Is it more important for a quarterback on a football team to be accurate or precise when throwing the football? Explain.

31. A student takes three mass measurements. The measurements have errors of 0.42 g, 0.38 g, and 0.47 g. What information would you need to determine whether these measurements are accurate or precise?

32. What conversion factor is needed to convert minutes to hours?

33. What kind of graph would you use to represent the following data?

    a. the segments of the population who plan to vote for a certain candidate

    b. the average monthly temperatures of two cities

    c. the amount of fat in three different kinds of potato chips

    d. the percent by mass of elements in Earth's atmosphere

    e. your scores on math quizzes during a year

    f. the effect of a hormone on tadpole growth

Matter—Properties and Changes

3.1 Properties of Matter

All of the material—the "stuff"—around us is matter. A **substance** is matter that has a uniform and unchanging composition. For example, water is a pure substance. No matter where it is found, a sample of water will have the same composition as any other sample of water.

A **physical property** of a substance is a characteristic that can be observed and measured without changing the composition of the substance. Words such as *hard*, *soft*, *shiny*, *dull*, *brittle*, *flexible*, *heavy* (in density), and *light* (in density) are used to describe physical properties.

A **chemical property** describes the ability of a substance to combine with or change into one or more other substances. For example, the ability of iron to form rust when combined with air is a chemical property of iron. The inability of a substance to combine with another substance is also a chemical property. For example, the inability to combine with most other substances is a chemical property of gold.

Practice Problems

1. Identify each of the following as an example of a physical property or a chemical property.

   a. Silver tarnishes when it comes in contact with hydrogen sulfide in the air.

   b. A sheet of copper can be pounded into a bowl.

   c. Barium melts at 725°C.

   d. Helium does not react with any other element.

   e. A bar of lead is more easily bent than is a bar of aluminum of the same size.

   f. Potassium metal is kept submerged in oil to prevent contact with oxygen or water.

g. Diamond dust can be used to cut or grind most other materials.

h. Rocks containing carbonates can be identified because they fizz when hydrochloric acid is applied.

Under ordinary conditions, matter exists in three different physical forms called the **states of matter**—solid, liquid, and gas. **Solid** matter has a definite shape and a definite volume. A solid is rigid and incompressible, so it keeps a certain shape and cannot be squeezed into a smaller volume. A solid has these properties because the particles that make up the solid are packed closely together and are held in a specific arrangement.

Liquid matter has a definite volume, like a solid, but flows and takes the shape of its container. A liquid is virtually incompressible because its particles are packed closely together. A liquid flows because the particles are held in no specific arrangement but are free to move past one another.

Like a liquid, a **gas** flows and takes the shape of its container, but has no definite volume and occupies the entire space of its container. Gaseous matter has these properties because its particles are free to move apart to fill the volume of the container. Also, because of the space between its particles, a gas can be compressed to a smaller volume. A **vapor** is the gaseous state of a substance that is a liquid or a solid at room temperature.

Practice Problems

2. Identify each of the following as a property of a solid, liquid, or gas. Some answers will include more than one state of matter.

a. flows and takes the shape of its container

b. compressible

c. made of particles held in a specific arrangement

d. has a definite volume

e. always occupies the entire space of its container

f. has a definite volume but flows

3.2 Changes in Matter

Matter can undergo two fundamental kinds of changes. Changes that do not alter the composition of matter are called **physical changes.** Phase changes, in which matter changes from one phase (or state) to another, are common examples of physical changes. The temperatures at which the phase changes of boiling and melting take place are important physical properties of substances.

A **chemical change** occurs when one or more substances change into new substances. A chemical change is also known as a **chemical reaction.** The appearance of new substances is the sign that a chemical reaction has occurred. In a chemical reaction, the substances present at the start are called **reactants.** The new substances that are formed in the reaction are called **products.** A chemical reaction is represented by a **chemical equation,** which shows the relationship between reactants and products.

Practice Problems

3. Identify each of the following as an example of a chemical change or a physical change.

   a. Moisture in the air forms beads of water on a cold windowpane.

   b. An electric current changes water into hydrogen and oxygen.

   c. Yeast cells in bread dough make carbon dioxide and ethanol from sugar.

   d. Olive oil, vinegar, salt, and pepper are shaken together to make salad dressing.

   e. Molten bronze is poured into a mold and solidifies to form a figurine.

   f. A reactant decomposes to form two products.

▶ **Conservation of Mass** Over two centuries ago, chemists established a fundamental law called the **law of conservation of mass.** This law states that during a chemical reaction, mass is neither lost nor gained. In other words, all the matter present at the start of a reaction still exists at the end of the reaction. The law of conservation of mass can be stated in mathematical form as follows.

$$Mass_{reactants} = Mass_{products}$$

Example Problem 3-1
Law of Conservation of Mass

A thin strip of iron with a mass of 15.72 g is placed into a solution containing 21.12 g of copper(II) sulfate and copper begins to form. After a while, the reaction stops because all of the copper(II) sulfate has reacted. The iron strip is found to have a mass of 8.33 g. The mass of copper formed is found to be 8.41 g. What mass of iron(II) sulfate has been formed in the reaction?

Solution Apply the law of conservation of mass.
In this reaction, there are two reactants and two products, so the law of conservation of mass can be restated as follows.

$$Mass_{reactant\ 1} + Mass_{reactant\ 2} = Mass_{product\ 1} + Mass_{product\ 2}$$

Rewrite the equation with the names of the reactants and products.

$$Mass_{iron} + Mass_{copper\ sulfate} = Mass_{copper} + Mass_{iron\ sulfate}$$

To find the mass of iron sulfate, rearrange the equation.

$$Mass_{iron\ sulfate} = Mass_{iron} + Mass_{copper\ sulfate} - Mass_{copper}$$

Then, determine the mass of iron that reacted.

$$Mass_{iron} = \text{original mass of iron} - \text{mass of iron remaining}$$

$$Mass_{iron} = 15.72\ g - 8.33\ g = 7.39\ g$$

Finally, substitute the masses into the equation and solve.

$$Mass_{iron\ sulfate} = 7.39\ g + 21.12\ g - 8.41\ g = 20.10\ g\ \text{iron sulfate}$$

To check your work, make sure the sum of the masses of the reactants is equal to the sum of the masses of the products.

Practice Problems

4. A sealed glass tube contains 2.25 g of copper and 3.32 g of sulfur. The mass of the tube and its contents is 18.48 g. Upon heating, a reaction forms copper(II) sulfide (CuS). All of the copper reacts, but only 1.14 g of sulfur reacts. Predict what the mass of the tube and its contents will be after the reaction is completed. Explain your reasoning.

5. When heated, calcium hydroxide and ammonium chloride react to produce ammonia gas, water vapor, and solid calcium chloride. Suppose 5.00 g of calcium hydroxide and 10.00 g of ammonium chloride are mixed in a test tube and heated until no more ammonia is given off. The remaining material in the test tube has a mass of 10.27 g. What total mass of ammonia and water vapor was produced in the reaction?

6. When a solution of barium nitrate and a solution of copper(II) sulfate are mixed, a chemical reaction produces solid barium sulfate, which sinks to the bottom, and a solution of copper(II) nitrate. Suppose some barium nitrate is dissolved in 120.00 g of water and 8.15 g of copper(II) sulfate is dissolved in 75.00 g of water. The solutions are poured together, and a white solid forms. After the solid is filtered off, it is found to have a mass of 10.76 g. The mass of the solution that passed through the filter is 204.44 g. What mass of barium nitrate was used in the reaction?

7. A reaction between sodium hydroxide and hydrogen chloride gas produces sodium chloride and water. A reaction of 22.85 g of sodium hydroxide with 20.82 g of hydrogen chloride gives off 10.29 g of water. What mass of sodium chloride is formed in the reaction?

3.3 Mixtures of Matter

A **mixture** is a combination of two or more pure substances in which each substance retains its individual properties. Concrete, most rocks, most metal objects, all food, and the air you breathe are mixtures that are often composed of many different substances. The composition of a mixture is variable. For example, the composition of salt water can be varied by changing the amount of salt or water in the mixture.

Two types of mixtures exist. A **heterogeneous mixture** is one that is not blended smoothly throughout. Examples of heterogeneous mixtures include smoky air and muddy water. You may have to use a magnifying glass or even a microscope, but if you can identify bits of one or more of the components of a mixture, the mixture is heterogeneous.

A **homogeneous mixture** is one that has a constant composition throughout. By dissolving sugar in water, you create a homogeneous mixture. A homogeneous mixture is also called a **solution.** In solutions, the atoms and/or molecules of two or more substances are completely mingled with one another. Solutions do not have to be solids dissolved in liquids; they can be mixtures of various states of matter. For example, air is a gaseous solution containing nitrogen, oxygen, argon, carbon dioxide, water vapor, and small amounts of other gases. An **alloy** is a homogeneous mixture (solution) of two or more metals or of metals and nonmetals. Alloys are considered to be solid solutions.

Practice Problems

8. Identify each of the following as an example of a homogeneous mixture or a heterogeneous mixture.
   a. 70% isopropyl rubbing alcohol
   b. a pile of rusty iron filings
   c. concrete
   d. saltwater
   e. gasoline
   f. bread

3.4 Elements and Compounds

Pure substances are classified into two categories—elements and compounds. An **element** is a substance that cannot be separated into simpler substances. Oxygen, carbon, sulfur, aluminum, iron, and gold are elements. More than 110 elements are known, but only 91 occur naturally on Earth. The remainder have been created by scientists.

Each element may be represented by a one- or two-letter symbol. In the mid-nineteenth century, a Russian chemist, Dmitri Mendeleev, saw that elements could be arranged according to repeating patterns of properties. Mendeleev's arrangement has evolved into the **periodic table** of the elements that you can see inside the back cover of this Handbook. The vertical columns of elements are called **groups** or **families.** The elements in a group have similar properties. Horizontal rows on the periodic table are called **periods.** When a new period starts, the elements tend to repeat the properties of the elements above them in the previous period. The value of Mendeleev's table was that it correctly predicted the properties of elements that had not yet been discovered.

A **compound** is a substance that consists of two or more elements chemically combined. The elements in a compound are not simply mingled together as they are in a mixture. Instead, they are combined in a way such that the compound has properties that are different from those of the elements of which it is composed. Each compound is represented by a formula, a combination of the symbols of the elements that make up the compounds. You may already know some chemical formulas such as H_2O for water, NaCl for table salt, and NH_3 for ammonia.

Practice Problems

9. Identify each of the following as an example of an element or a compound.

   a. sucrose (table sugar)

   b. the helium in a balloon

   c. baking soda

   d. a diamond

   e. aluminum foil

   f. the substances listed on the periodic table

   g. calcium chloride pellets used to melt ice

A given compound is always composed of the same elements in the same proportion by mass. This fact is known as the **law of definite proportions.** For example, 100.00 g of H_2O always contains 11.19 g of hydrogen and 88.81 g of oxygen, no matter where the water came from. Compounds are often identifiable from their percentage composition, the **percent by mass** of each element in a compound. The percent by mass of each element in a compound may be found by using the following equation.

$$\text{Percent by mass of an element (\%)} = \frac{\text{mass of element}}{\text{mass of compound}} \times 100\%$$

For example, suppose you break a compound down into its elements and find that 25.00 g of the compound is composed of 6.77 g of tin and 18.23 g of bromine. The percent by mass of tin in the compound can be determined as follows.

$$\text{Percent by mass of tin} = \frac{\text{mass of tin}}{\text{mass of compound}} \times 100\%$$

$$\text{Percent by mass of tin} = \frac{6.77 \text{ g}}{25.00 \text{ g}} \times 100\% = 27.1\% \text{ tin}$$

Practice Problems

10. Follow the procedure described above to determine the percent by mass of bromine in the compound discussed above. What is the sum of the percents of the two elements?

11. A 134.50-g sample of aspirin is made up of 6.03 g of hydrogen, 80.70 g of carbon, and 47.77 g of oxygen. What is the percent by mass of each element in aspirin?

12. A 2.89-g sample of sulfur reacts with 5.72 g of copper to form a black compound. What is the percentage composition of the compound?

13. Aluminum oxide has a composition of 52.9% aluminum and 47.1% oxygen by mass. If 16.4 g of aluminum reacts with oxygen to form aluminum oxide, what mass of oxygen reacts?

Chapter 3 Review

14. How does a pure substance differ from a mixture of substances such as lemonade?

15. Identify each of the following as an example of a physical property or a chemical property.

   a. A piece of silver can be hammered into a cloverleaf shape.

   b. A piece of charcoal, which is mostly the substance carbon, glows red, gives off heat, and becomes a gray ash.

   c. Table salt dissolves in water and remains as a crust when the water evaporates.

   d. The Statue of Liberty, once the color of a new penny, turned to a bluish-green as it formed a thin coating called a patina.

16. Can two different samples of the same substance have different physical properties? Can they have different chemical properties? Explain your answers to both questions.

17. Write one statement each for a solid, a liquid, and a gas describing the particle arrangement and spacing in each state.

18. Suppose you measure the mass of an iron nail and find it to be 13.8 g. You place the nail in a moist place for two weeks. When you retrieve the nail, it is covered with a crusty brown coating of rust. You then place the nail on a balance to measure its mass. Will the nail have a mass that is greater than, less than, or equal to 13.8 g? Explain your answer.

19. Identify each of the following examples as a physical change or a chemical change.

   a. Wood burns in a fireplace.

   b. A block of pewter is pounded and made into a bowl.

   c. A copper weather vane becomes green in a few years.

   d. Baking soda in cookie dough causes the cookies to puff with bubbles of gas when they are baked.

   e. A pan of water boils on a stove until the pan becomes dry.

   f. Soap and water are used to clean up a spill of grease.

   g. Mothballs vaporize in a closed closet.

20. Aluminum can combine chemically with various substances, but it never reacts to form simpler substances. Is aluminum an element, a compound, or a mixture? How do you know? Is aluminum a substance? Explain.

21. The mineral cassiterite is mined as a commercial source of tin. It is a compound of the elements tin and oxygen. A 1.000-kg sample of pure cassiterite mined in Germany contains 788 g of tin. Is it possible to predict the mass of tin contained in a 43.5-kg sample of cassiterite mined in South America? Explain your reasoning.

22. The elements carbon and sulfur are solids at room temperature. Can you reasonably predict that a compound of these two elements will also be a solid at room temperature? Explain your conclusion.

The Structure of the Atom

4.1 Early Theories of Matter

In ancient Greece, philosophers debated the nature of matter. The philosopher Democritus held that matter was composed of elementary particles, called **atoms,** in otherwise empty space. Over two thousand years would pass before the existence of the atoms was proven.

▶ **The atomic theory** In 1803, John Dalton proposed a theory to explain the laws of conservation of matter, definite proportions, and multiple proportions. **Dalton's atomic theory,** based on experimental results, held that matter consists of atoms, that all atoms of an element are alike, and that atoms of one element differ from atoms of other elements. Further, the theory proposed that atoms were indestructible and only were rearranged during chemical reactions.

4.2 Subatomic Particles and the Nuclear Atom

One of the most important results of Dalton's theory was that it generally persuaded scientists that atoms exist. As a result, scientists began research to discover the exact nature of the atom.

▶ **Discovery of electrons** Experiments around the turn of the twentieth century provided evidence that the atom consists of smaller particles. When J. J. Thomson applied a high voltage to two electrodes sealed inside an evacuated tube, an invisible beam, or ray, was found to emanate from the negative electrode (cathode) of the tube and travel toward the positive electrode (anode). The path of the beam could be made visible by the glow it caused when it passed across a plate coated with fluorescent material. The ray, called a **cathode ray,** could be deflected by electric and magnetic fields.

Thomson concluded that the ray consisted of a stream of negatively charged particles that had been dislodged from atoms. These particles became known as **electrons.** Thomson went on to determine that the mass of an electron is much less than that of a hydrogen atom.

Later, in 1909, Robert Millikan accurately determined the mass of an electron to be 1/1840 the mass of a hydrogen atom. He also determined the electron's charge to be $1-$. This value was accepted as a fundamental unit of electrical charge.

▶ **The nucleus of an atom** Because atoms have no net electrical charge, scientists reasoned that if an atom contains particles of negative charge, it must also contain particles of positive charge. In 1911, Ernest Rutherford conducted experiments in which he aimed a beam of alpha particles (positively charged helium nuclei) at a thin sheet of gold. The fact that a few of these heavy particles were deflected almost directly backwards revealed that the positive charge of an atom, along with nearly all of the atom's mass, is concentrated in a very small region in the center of the atom. This positively charged mass is called the **nucleus** of the atom. The particles that make up the positive charge of the nucleus are called **protons,** and each has a charge of $1+$ (equal to but opposite the charge of an electron).

In 1932, Chadwick demonstrated the existence of the **neutron,** a nuclear particle having no charge but with nearly the same mass as a proton. The following table summarizes the subatomic particles that comprise the atom.

Properties of Subatomic Particles				
Particle	**Symbol**	**Electrical charge**	**Relative mass**	**Actual mass (g)**
Electron	e^-	$1-$	1/1840	9.11×10^{-28}
Proton	p^+	$1+$	1	1.673×10^{-24}
Neutron	n^0	0	1	1.675×10^{-24}

Practice Problems

1. Select the term in Column B that best matches the phrase in Column A.

Column A	Column B
a. Cathode ray	**1.** electron
b. Discovered in 1932	**2.** proton
c. Caused large deflections of alpha particles in Rutherford's experiment	**3.** neutron
d. Has a charge of $1-$	**4.** nucleus
e. Has no charge	
f. Contains nearly all of an atom's mass	
g. In an atom, the number of these particles is equal to the number of protons.	
h. Identified by Thomson	
i. Site of an atom's positive charge	
j. Has a positive charge and relative mass of 1	
k. The center of an atom	
l. Symbolized by n^0	

4.3 How Atoms Differ

It is important to be able to obtain useful information from the periodic table. Starting with the simplest element, hydrogen, the periodic table represents each element in a box.

▶ **Information about an element** The name of the element and its chemical symbol are given in each box. The integer in the box represents the element's **atomic number,** the number of protons in a nucleus of an atom of that element. As you scan the table, you can see that the elements are arranged in order of increasing atomic number and that the sequence of atomic numbers has no missing values. This continuous series is an indication of the fact that the atomic number determines the identity of an element.

Atoms have no net electrical charge because they contain equal numbers of electrons and protons. Thus, positive charges (protons) and negative charges (electrons) balance out. Electrons, protons, and atomic number are related as follows.

Atomic number = number of protons = number of electrons

Practice Problems

2. How many electrons and protons are contained in an atom of each of the following elements?

   a. arsenic   **d.** molybdenum

   b. gold   **e.** polonium

   c. fluorine   **f.** barium

3. Identify the atom containing the following number of electrons.

   a. 34   **d.** 61

   b. 5   **e.** 94

   c. 31

4. Identify the atom containing the following number of protons.

   a. 74   **d.** 70

   b. 20   **e.** 93

   c. 49

5. Atoms of which element contain 18 protons and 18 electrons?

▶ **Isotopes and mass number** Although all atoms of an element have the same number of protons, most elements have atoms with different numbers of neutrons. For example, because the atomic number of hydrogen is one, all atoms of hydrogen must have one proton (and one electron). However, hydrogen atoms can have different numbers of neutrons. Approximately 99 985 out of every 100 000 naturally occurring hydrogen atoms have no neutrons. However, the remaining 15 out of every 100 000 hydrogen atoms have one neutron. Atoms with the same number of protons but different numbers of neutrons are called **isotopes.** Hydrogen has two naturally occurring isotopes.

An isotope of an element is identified by its **mass number,** which is the sum of the number of protons and neutrons in the atom's nucleus.

Mass number = number of protons + number of neutrons

When writing the symbol for an isotope, write the atomic number and mass number as shown in the following examples.

H atom with one proton
and no electrons

H atom with one proton
and one neutron

Mass number $\longrightarrow$ 1
Atomic number $\longrightarrow$ 1 H

Mass number $\longrightarrow$ 2
Atomic number $\longrightarrow$ 1 H

Isotopes of an element can also be specified by writing the mass number of the isotope following a hyphen after the name of the element. For example, hydrogen's isotopes can be written as hydrogen-1 and hydrogen-2, or H-1 and H-2, respectively.

Recall that the number of protons in an atom's nucleus is its atomic number. Substituting atomic number for the number of protons gives the following relationship.

Number of neutrons = mass number − atomic number

You can use this relationship to determine any one of the variables if you know the other two.

Example Problem 4-1
Using Atomic Number and Mass Number

One of the four naturally occurring isotopes of chromium has a mass number of 53. Determine the number of protons, electrons, and neutrons in an atom of this isotope and write its symbol.

Start by obtaining information on chromium from the periodic table. Chromium (Cr) is found in the fourth row of the table in group 6. The atomic number of chromium is 24.

You know that the atomic number of an element gives the number of protons in the nucleus of the atom.

Atomic number = number of protons = 24

You have also learned that a neutral atom has equal numbers of protons and electrons, so you can write the following expression.

Number of electrons = number of protons = 24

You also know that the number of neutrons can be determined from the mass number and the atomic number.

Number of neutrons = mass number − number of protons

Number of neutrons = 53 − 24 = 29

To write the symbol of this isotope, follow this pattern.

$$\text{Mass number} \longrightarrow {}^{a}_{b}X \longleftarrow \text{Symbol of element}$$
$$\text{Atomic number} \longrightarrow$$

Substituting the known values gives this symbol for the isotope.

$$^{53}_{24}\text{Cr}$$

Practice Problems

6. The other three naturally occurring isotopes of chromium have mass numbers of 50, 52, and 54. Describe how atoms of these isotopes differ from the isotope mentioned in Example Problem 4-1.

7. All naturally occurring atoms of fluorine have a mass number of 19. Determine the number of protons, electrons, and neutrons in an atom of fluorine and write the atom's symbol.

8. Describe the subatomic particles comprising an isotope of zirconium-94 ($^{94}_{40}$Zr).

9. An atom of a certain element has a mass number of 112 and is known to contain 64 neutrons. Identify the atom and determine the number of electrons and protons the atom contains.

10. A neutral atom has 78 electrons and a mass number of 198. Identify the atom and determine the number of protons and neutrons in its nucleus.

▶ **Atomic mass** Scientists have established a standard for the measurement of atomic mass by assigning the carbon-12 atom a mass of 12 **atomic mass units (amu).** Thus, 1 amu is equal to 1/12 the mass of a carbon-12 atom.

The number at the bottom of each square in the periodic table is the atomic mass of that element in amu. A scan of the periodic table will immediately tell you that the atomic masses of many elements are not whole numbers. The reason for this is that many elements occur in nature as a mixture of isotopes—isotopes with different masses. The **atomic mass** of an element given in the table is a weighted average of the atomic masses of the naturally occurring isotopes of the element. This weighted average takes into account the mass and abundance of each of the isotopes. The following Example Problem shows you how to calculate the atomic mass of an element.

Example Problem 4-2
Calculating Atomic Mass

Copper exists as a mixture of two isotopes. The lighter isotope (Cu-63), with 29 protons and 34 neutrons, makes up 69.17% of copper atoms. The heavier isotope (Cu-65), with 29 protons and 36 neutrons, constitutes the remaining 30.83% of copper atoms. The atomic mass of Cu-63 is 62.930 amu, and the atomic mass of Cu-65 is 64.928 amu. Use the data above to compute the atomic mass of copper.

For a problem that has several numerical values, it is wise to organize the data into a table.

Abundance and Mass Data for Copper		
	Isotope	
	Copper-63	**Copper-65**
Number of protons	29	29
Number of neutrons	34	36
Atomic mass	62.930 amu	64.928 amu
Abundance	69.17%	30.83%

First, calculate the contribution of each isotope to the average atomic mass, being sure to convert each percent to a fractional abundance.

Mass contribution = mass of isotope × abundance of isotope

For Cu-63:

Mass contribution = 62.930 amu × 0.6917 = 43.529 amu

For Cu-65:

Mass contribution = 64.928 amu × 0.3083 = 20.017 amu

The average atomic mass of the element is the sum of the mass contributions of each isotope.

Atomic mass Cu = mass contribution Cu-63
+ mass contribution Cu-65

Atomic mass Cu = 43.529 amu + 20.017 amu = 63.546 amu

You can easily confirm this result by checking the periodic table.

Practice Problems

11. Gallium occurs in nature as a mixture of two isotopes. They are Ga-69 with a 60.108% abundance and a mass of 68.926 amu and Ga-71 with a 39.892% abundance and an atomic mass of 70.925. Calculate the atomic mass of gallium.

12. The following table shows the five isotopes of germanium found in nature, the abundance of each isotope, and the atomic mass of each isotope. Calculate the atomic mass of germanium.

Isotope	Abundance (%)	Atomic Mass (amu)
Germanium-70	21.23	69.924
Germanium-72	27.66	71.922
Germanium-73	7.73	72.923
Germanium-74	35.94	73.921
Germanium-76	7.44	75.921

13. The atomic mass of bromine given in the periodic table is 79.904 amu, which is very close to 80 amu. Use a reference book to find the percent of Br-80 in naturally occurring bromine. Explain the value of the atomic mass of bromine from the data you find.

14. The element chlorine occurs in nature as a mixture of two isotopes. Chlorine-35 has an atomic mass of 34.969 amu and makes up 75.77% of chlorine atoms. Chlorine-37 atoms make up the remaining 24.23% of all chlorine. Use the average atomic mass of chlorine from the periodic table to calculate the atomic mass of Cl-37 atoms.

4.4 Unstable Nuclei and Radioactive Decay

Dalton thought that atoms were unchangeable. However, scientists in the late nineteenth century began to notice that some elements spontaneously emitted energy and particles. These emissions became known as **radiation.** Elements that give off radiation are said to be **radioactive.** Soon, scientists discovered that radioactive elements changed into other elements after emitting radiation. Thus, atoms were not unchangeable as Dalton had thought.

Today, we know that some nuclei are unstable and gain stability by emitting radiation. This process is called **radioactive decay.**

▶ **Types of radiation** Three principal types of radiation are given off during radioactive decay. One type is **alpha radiation** (α-radiation), which consists of **alpha particles** ejected from the nucleus. An alpha particle is equivalent to a helium-4 nucleus, that is, a nucleus with two protons and two neutrons. Alpha particles have a $2+$ charge.

$$\alpha\text{-particle: } {}^4_2\text{He (contains } 2p^+ \text{ and } 2n^0)$$

The second common type of radiation is **beta radiation,** which consists of high-energy electrons called **beta particles** (β-particles).

$$\beta\text{-particle: } {}^0_{-1}\beta \text{ (high-energy } e^-)$$

It is important to realize that each of these beta particles is emitted by the atom's nucleus, thereby increasing its atomic number (and charge) by $1+$.

The third type of radiation is **gamma radiation** (γ-radiation), which is high-energy electromagnetic radiation. Gamma radiation is usually emitted along with α-particles and β-particles.

Chapter 4 Review

15. Why were Dalton's ideas about atoms considered a scientific theory, whereas the ideas of ancient Greek philosophers were not?

16. Why was the discovery of the electron considered evidence for the existence of positively charged particles in the atom?

17. What led scientists to conclude that most of the atom's mass is concentrated in a very small nucleus? What particles are responsible for that mass?

18. Explain the relationship between the number of protons and the number of electrons in an atom of argon.

19. Can atoms of two different elements have the same number of protons? Explain your answer.

20. What particles make up an atom of selenium-80?

21. Explain the meaning of the number printed at the bottom of each box of the periodic table.

22. Write symbols for the following isotopes.
   a. antimony-123
   b. silicon-28
   c. mercury-199

23. Complete the following table.

Types of Radiation from Radioactive Elements			
Type of radiation	**Mass number**	**Charge**	**Composition**
	1/1840		electron
		2+	
Gamma			energy only

Electrons in Atoms

5.1 Light and Quantized Energy

In the early twentieth century, scientists accepted the idea that an atom consisted of a massive, positively charged nucleus surrounded by negatively charged electrons. Further explanation of the atom's electron arrangement came from research involving light and its interaction with matter.

▶ **Wave nature of light** Visible light is part of a range of **electro-magnetic radiation,** a form of wave energy that travels through empty space and is propagated in the form of alternating electric and magnetic fields. X rays, gamma rays, and radio waves are other examples of electromagnetic radiation.

Electromagnetic waves, along with water waves and sound waves, exhibit certain common characteristics. All waves consist of a series of crests and troughs that travel away from their source at a velocity that is determined by the nature of the wave and the material through which the wave passes. The rate of vibration of a wave is called its **frequency** and is defined as the number of waves that pass a given point per second. Wave frequency is expressed in hertz (Hz); one hertz equals one wave per second (s^{-1}).

The frequency and velocity of a wave determine its **wavelength,** the distance between equivalent points on a continuous wave. For waves of a given velocity, a wave with a higher frequency has a shorter wavelength. For electromagnetic waves, this inverse relation is expressed mathematically by the following equation.

$$c = \lambda \nu$$

In this equation, c equals 3.00×10^8 m/s, the velocity of electromagnetic waves in a vacuum; λ equals the wavelength in meters; and ν is the frequency of the waves in hertz.

Practice Problems

1. A helium-neon laser emits light with a wavelength of 633 nm. What is the frequency of this light?

2. What is the wavelength of X rays having a frequency of 4.80×10^{17} Hz?

3. An FM radio station broadcasts at a frequency of 98.5 MHz. What is the wavelength of the station's broadcast signal?

▶ **Particle nature of light** One of the first major clues to the electron structure of atoms was Albert Einstein's explanation of the **photoelectric effect,** a phenomenon in which electrons are ejected from the surface of a polished metal plate when it is struck by light. Whether or not electrons are ejected depends on the frequency (color) of the incident light. In 1905, Einstein reasoned that this phenomenon could be explained only if light could act like a stream of particles that knocked electrons out of atoms. Particles, or **photons,** of light at the high-frequency (violet) end of the visible spectrum had greater energy and were therefore more effective at dislodging electrons.

The energy of a photon (E_{photon}) of a certain frequency (ν) can be calculated by using the following equation in which h represents **Planck's constant** ($h = 6.626 \times 10^{-34}$ J·s).

$$E_{photon} = h\nu$$

Practice Problems

4. Calculate the energy of a gamma ray photon whose frequency is 5.02×10^{20} Hz.

5. What is the difference in energy between a photon of violet light with a frequency of 6.8×10^{14} Hz and a photon of red light with a frequency of 4.3×10^{14} Hz?

6. Calculate the energy of a photon of ultraviolet light that has a wavelength of 49.0 nm.

▶ **Atomic emission spectra** When atoms of an element in the gaseous phase are excited by an input of energy, such as that from a high-voltage electrical source, they emit light. (A neon sign works on this principle.) This emitted light can be broken into a spectrum consisting of discrete lines of specific frequencies, or colors. This pattern of frequencies (colors) is unique to each element and is known as the element's **atomic emission spectrum.**

Consider the fact that each discrete frequency consists of

photons of a specific energy value. An element's atomic emission spectrum, therefore, is evidence that an atom can emit light of only certain specific energies. The explanation of this phenomenon unlocked the secret of the atom's electron arrangement.

5.2 Quantum Theory and the Atom

The emission spectrum of an atom indicates that the energy it emits is quantized, meaning that only certain quantities of energy can be given off. Niels Bohr used the relatively simple emission spectrum of the hydrogen atom to propose a new atomic model.

▶ **The Bohr atomic model** Bohr proposed that hydrogen's single electron could orbit at only specific distances from the atom's nucleus. When the electron is in the orbit nearest the nucleus, it has the lowest possible energy. When the electron is in the next larger orbit, it has a higher energy, and so on through the larger allowed orbits. The electron can occupy only those specific allowed orbits and thus can have only the energies associated with those orbits.

When an electron is excited by an outside input of energy, it can absorb only an amount of energy needed to jump to one of the higher-energy orbits. When it falls back to a lower-energy orbit, the electron emits an amount (quantum) of energy equal to the difference in energy between the two orbits. Because the emission spectrum of hydrogen consisted of several different frequencies, Bohr assumed that the electron could occupy orbits of several different energies, designated by integers called quantum numbers.

Bohr's model of the atom, sometimes called the planetary model, was accepted at the time because it explained hydrogen's atomic emission spectrum. However, the model was very limited in that it only worked for hydrogen.

▶ **De Broglie's waves** Recall that Einstein explained the photoelectric effect by reasoning that electromagnetic waves could act like particles. In 1924, Louis de Broglie turned the tables by theorizing that particles of matter, specifically electrons, could exhibit the properties of waves—frequency and wavelength. De Broglie proposed that electrons of only a specific frequency could fit into one of the possible atomic orbits.

▶ **The modern atomic model** De Broglie's work, followed by the work of Werner Heisenberg and Irwin Schrödinger, led to the modern **quantum mechanical model of the atom.** This model may be summarized as follows.

1. Electrons occupy the space surrounding the nucleus and can exist in several discrete **principal energy levels,** each designated by one of the **principal quantum numbers** (n) that are the integers 1, 2, 3, 4, and so on.

2. Electrons in successively higher principal energy levels have greater energy.

3. Because of interactions among electrons, each principal energy level consists of **energy sublevels** that have slightly different energy values. These sublevels are designated by the letters s, p, d, and f in order of increasing energy. The first principal energy level ($n = 1$) has only one sublevel (s). The second principal energy level ($n = 2$) consists of two sublevels (s and p). The third principal energy level consists of three sublevels (s, p, and d), and the fourth consists of four sublevels (s, p, d, and f).

4. Each energy sublevel consists of one or more orbitals, each of which can contain two electrons. An s sublevel has one orbital; a p sublevel has three orbitals; a d sublevel has five orbitals; and an f sublevel has seven orbitals. All of the orbitals in the same sublevel are of equal energy.

5. Atomic orbitals are regions of space in which there is a high probability (90 percent) of finding an electron. Except for s orbitals, the orbitals are not spherical in shape. Unlike Bohr's atomic model, electrons are not in specifically defined orbits. Instead, they can be anywhere within the orbital, and there is a 10 percent chance that the electron will be located outside the orbital.

Practice Problems

7. What feature of Bohr's atomic model accounted for the fact that electrons can have only certain energies?

8. Describe what is happening when an atom emits a photon.

9. How many electrons can the second principal energy level hold? How many electrons can the third principal energy level hold? Explain the difference in these numbers of electrons.

5.3 Electron Configurations

As you will learn, the number and arrangement of electrons around the nucleus of an atom determines its chemical properties. Therefore, it is useful to be able to determine and write out the electron arrangement, called an **electron configuration,** of an atom in symbolic form.

▶ **Writing electron configurations** The electron configuration of an atom is written by stating the number of electrons in each energy sublevel and writing the sublevels in order of increasing energy. The number of electrons in an energy sublevel is indicated by a superscript integer. For example, the electron configuration of a hydrogen atom, which has one electron, is written as $1s^1$. This indicates that hydrogen's one electron is in the 1s sublevel.

Helium has two electrons. Recall that an s sublevel consists of a single orbital that can hold a maximum of two electrons. Therefore, helium's second electron fills the one available orbital in the 1s sublevel, and its electron configuration is written as $1s^2$.

A neutral lithium atom has three electrons. The first principal energy level is filled with two electrons. Where does the third electron go? This electron is found in the second principal energy level, which, like all energy levels, begins with an s sublevel. So, lithium has two electrons in the 1s orbital and a third electron in the 2s orbital, giving it the electron configuration $1s^2 2s^1$. With beryllium, which has four electrons, the 2s orbital is filled with two electrons, yielding a configuration of $1s^2 2s^2$.

The element boron has five electrons, four of which fall into the same configuration as beryllium. Recall that the second principal energy level has two energy sublevels available (s and p) and that the p sublevel is of higher energy than the s sublevel. One rule governing electron configurations is the **aufbau principle,** which states that each successive electron occupies the lowest energy orbital available. In the case of boron, the lower-energy s orbitals are full, so the fifth electron is found in one of the three available orbitals in the higher-energy 2p sublevel. Therefore, boron's electron configuration is $1s^22s^22p^1$.

Recall that the three orbitals available in a p sublevel can each hold a maximum of two electrons. Continuing from boron through carbon ($1s^22s^22p^2$), nitrogen ($1s^22s^22p^3$), oxygen ($1s^22s^22p^4$), fluorine ($1s^22s^22p^5$), and neon ($1s^22s^22p^6$), the 2p sublevel becomes filled with six electrons.

With the element sodium, the eleventh electron begins the 3s sublevel to give the configuration $1s^22s^22p^63s^1$. The same pattern that occurred with lithium through neon repeats here, with successive electrons filling the 3s orbital and 3p orbitals. The element argon has a filled 3p sublevel. Argon has 18 total electrons, and the configuration $1s^22s^22p^63s^23p^6$.

You may recall that three sublevels—s, p, and d—are available in the third energy level, so you might expect the next electron to begin the 3d sublevel. However, a complication occurs here because the 4s sublevel is of *lower* energy than the 3d sublevel. Thus, following the aufbau principle, the next (nineteenth) electron begins the 4s sublevel in the element potassium, producing the configuration $1s^22s^22p^63s^23p^64s^1$. Scandium is the first element that has electrons in the 3d sublevel. As an example of an element that has electrons in the 4p sublevel, the electron configuration of arsenic is $1s^22s^22p^63s^23p^64s^23d^{10}4p^3$. Notice that the configuration is written by placing the sublevels in order of increasing energy, *not* in numerical order.

Practice Problems

10. When writing the electron configuration of an atom, in what general order are the sublevels written?

11. How is the number of electrons in an energy sublevel indicated in an electron configuration?

12. Write the electron configurations of the following elements.

a. sulfur

b. calcium

c. bromine

d. magnesium

To place energy sublevels in order of increasing energy, it is useful to learn the following energy sublevel diagram. By following the arrows, you can write the energy sublevels in the correct order.

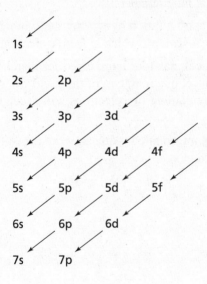

Because a new principal energy level always begins with the element immediately following one of the noble gases, it is possible to simplify electron configurations by using the symbol of the previous noble gas to denote all of an atom's inner-level electrons. As a simple example, consider nitrogen's electron configuration of $1s^2 2s^2 2p^3$. Helium, the preceding noble gas, has the configuration $1s^2$. Thus, the symbol [He] can be substituted into nitrogen's configuration to give $[He]2s^1 2p^3$. This form of an electron configuration is called noble-gas notation.

Example Problem 5-1
Writing Electron Configurations Using Noble-Gas Notation

Use noble-gas notation to write the electron configuration for a neutral atom of silicon.

To begin, find silicon on the periodic table and locate the preceding noble gas. Silicon (Si) is atomic number 14, and the preceding noble gas is neon (Ne), atomic number 10.

Start the electron configuration by writing the symbol [Ne]. Note that [Ne] replaces $1s^2 2s^2 2p^6$, which is neon's electron configuration.

Next, note that silicon has four more electrons than neon. According to the energy sublevel diagram, two of these electrons will fill the 3s sublevel, and the remaining two electrons will occupy orbitals in the 3p sublevel.

By combining this information, you can write the following electron configuration for silicon.

$$\text{Si} \quad [Ne]3s^2 3p^2$$

Practice Problems

13. Using noble-gas notation, write the electron configurations of the following elements.

a. fluorine

b. phosphorus

c. calcium

d. cobalt

e. selenium

f. technetium

g. iodine

h. holmium

i. iridium

j. radium

▶ **Valence electrons** When elements combine chemically, only the electrons in the highest principal energy level of each atom are involved. Therefore, these outermost electrons, called **valence electrons,** determine most of the chemical properties of an element. Later in your chemistry course, you will study the way in which elements form chemical bonds. Because bonding involves an atom's valence electrons, it is useful to be able to sketch a representation of an element's valence electrons. The American chemist G. N. Lewis devised the electron-dot structure to show an atom's valence electrons by writing dots around the symbol of the element.

In writing electron-dot structures, a single dot is used to represent each valence electron. One dot is placed on each of the four sides around the symbol before any two dots are paired together. The following examples illustrate the process. The electron-dot structure for hydrogen, which has one electron, is $H\cdot$. The electron-dot structure for helium, $1s^2$, is $\cdot He\cdot$. The electron configuration of lithium is $1s^2 2s^1$, but the two 1s electrons are in a stable inner energy level and do not participate in chemical changes. Only the outermost 2s electron is a valence electron, so the electron-dot structure for lithium is $Li\cdot$. The electron configuration of beryllium is $1s^22s^2$, but only the two 2s electrons are valence electrons, so the electron-dot structure for beryllium is •Be•. The electron-dot structure for boron, with three electrons in the second energy level ($1s^22s^22p^1$), is •Ḃ•. The electron configuration of oxygen is $1s^22s^22p^4$, and its electron-dot structure is •Ö:. A new principal energy level begins with sodium, whose electron configuration is $1s^22s^22p^63s^1$. Sodium's electron-dot structure is Na• because only the 3s electron is a valence electron. The other electrons are in inner energy levels.

Example Problem 5-2
Writing Electron-Dot Structures

Write an electron-dot structure for a neutral atom of arsenic, an element used in semiconductor materials because of its electron configuration.

Begin by writing the electron configuration of arsenic. Use noble-gas notation to emphasize the arrangement of electrons in the highest principal energy level. Arsenic is in the fourth period of the periodic table, so the preceding noble gas is argon. Thus, arsenic's electron configuration will begin with the symbol [Ar]. Arsenic has 15 electrons more than argon.

Using the energy sublevel diagram, you can see that the nineteenth electron begins the fourth principal energy level. After the 4s sublevel is filled with two electrons, ten electrons fill the 3d sublevel, which is slightly higher in energy than the 4s sublevel. After the 3d sublevel is filled, the remaining three electrons will be in orbitals in the 4p sublevel. Thus, the electron configuration for arsenic is $[Ar]4s^23d^{10}4p^3$.

Before writing the electron-dot structure for arsenic, note that arsenic's ten 3d electrons are not in the highest principal energy level. Therefore, they are not regarded as valence electrons. Only the 4s and 4p electrons are valence electrons; thus, the electron-dot structure of arsenic has five dots as follows.

•Aṡ:

Practice Problems

14. Write electron-dot structures for the following elements.

    a. nitrogen

    b. aluminum

    c. neon

    d. strontium

    e. antimony

    f. iodine

    g. lead

    h. cesium

15. What electron-dot structure is shared by all noble gases except helium?

16. List three elements that have the electron-dot structure $\cdot\overset{\cdot}{\underset{\cdot}{X}}:$.

Chapter 5 Review

17. When light passes from air into a denser material, such as glass or water, it slows down. How will the light's wavelength change as it slows down? Explain.

18. Describe the photoelectric effect. Based on the photoelectric effect, what did Einstein conclude about the nature of light?

19. How does the energy of a photon of electromagnetic energy change as the frequency increases? As the wavelength increases?

20. How did Bohr's atomic model explain the fact that the atomic emission spectrum of an element consists of lines of only certain colors?

21. How many energy sublevels are available in the third principal energy level? How many electrons can each of these sublevels hold?

22. In what significant way does the modern model of the atom differ from Bohr's model?

23. Identify the elements that have the following electron configurations.

   a. $[He]2s^22p^5$

   b. $[Ar]4s^23d^5$

   c. $[Ar]4s^23d^{10}4p^2$

   d. $[Ne]3s^1$

   e. $[Kr]5s^14d^8$

   f. $[Xe]6s^24f^{14}5d^{10}6p^4$

   g. $1s^22s^22p^63s^23p^4$

   h. $1s^22s^22p^63s^23p^64s^23d^{10}4p^65s^24d^{10}$

24. What electrons do the dots in an electron-dot structure represent? Why are these electrons important?

25. List three elements that have the electron-dot structure $\cdot X \cdot$.

The Periodic Table and Periodic Law

6.1 Development of the Modern Periodic Table

In 1869, the Russian chemist Dmitri Mendeleev noted that when the known elements were placed in order of increasing atomic mass, their properties repeated in a regular pattern—a periodic pattern. Mendeleev made a table in which he arranged the elements in order of increasing atomic mass into columns with similar properties. From his table, Mendeleev predicted the properties of three elements that were undiscovered at the time. When these elements—scandium, gallium, and germanium—were soon discovered and found to have most of the predicted properties, Mendeleev's periodic table was accepted in the scientific world. Besides the addition of newly discovered elements, the only major change in the organization of the periodic table since Mendeleev's time is that the elements are now arranged in order of atomic number instead of atomic mass.

▶ **The modern periodic table** Study the modern periodic table that appears in your textbook on pages 156–157. Each square gives certain information about each element, as shown in the following diagram.

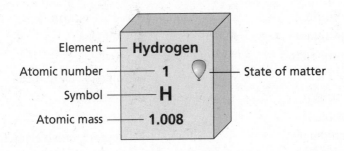

The elements are arranged in the periodic table in order of increasing atomic number in horizontal rows called **periods.** Because the pattern of properties repeats in each new row of elements, the elements in a column have similar properties and are called a **group** or family of elements. The groups are designated with a number and the letter A or B. Groups 1A through 8A are called the main group or **representative elements.** The group B elements are called the **transition elements.**

Elements are divided into three main classes—metals, metalloids, and nonmetals. As you can see from the periodic table, the majority of the elements are metals. **Metals** are generally shiny solids and are good conductors of heat and electrical current. Some groups of elements have names. For example, the first two groups of metals, groups 1A and 2A, are called the **alkali metals** and the **alkaline earth metals,** respectively. Most of the elements to the right of the heavy stair-step line in the periodic table are **nonmetals,** which are generally either gases or brittle solids at room temperature. Group 7A elements are commonly called **halogens,** and group 8A elements are the **noble gases.** Many of the elements that border the stair-step line are **metalloids,** which have some of the characteristics of both metals and nonmetals.

Practice Problems

Match each element in Column A with the best matching description from Column B. Each Column A element may match more than one description from Column B.

Column A	Column B
1. strontium	**a.** halogen
2. chromium	**b.** noble gas
3. iodine	**c.** alkaline earth metal
4. nitrogen	**d.** metalloid
5. argon	**e.** alkali metal
6. rubidium	**f.** representative element
7. silicon	**g.** transition element

6.2 Classification of the Elements

Chemists today understand that the repetition of properties of elements occurs because the electron configurations of atoms exhibit repeating patterns. Thus, the arrangement of elements in the periodic table reflects the electron structures of atoms.

For example, the group number of a representative element gives the number of valence electrons in an atom of that element. The group 3A element aluminum, for example, has the electron configuration $1s^22s^22p^63s^23p^1$, or $[Ne]3s^23p^1$. The three electrons in the third energy level ($3s^2$ and $3p^1$) are the valence electrons of the aluminum atom. In a similar way, the period number of a representative element indicates the energy level of the valence electrons. Aluminum is in the third period, and aluminum's valence electrons are in the third energy level.

Practice Problems

Use the periodic table to answer the following questions.

8. How many valence electrons are in an atom of each of the following elements?
   a. magnesium   d. arsenic
   b. selenium   e. iodine
   c. tin

9. In which energy level are the valence electrons of the elements listed in question 8?

10. Identify each of the following elements.
    a. an electron configuration of $[Kr]5s^24d^{10}5p^2$
    b. five valence electrons in the sixth energy level
    c. two valence electrons in the first energy level
    d. three fewer electrons in the fourth energy level than krypton
    e. an electron configuration ending in $4p^2$

▶ **Blocks of elements** The periodic table is divided into blocks of elements that correspond to the energy sublevel being filled as you move across a period. The 1A and 2A groups constitute the s-block elements because their highest-energy electrons are in s orbitals. The remaining groups of representative elements, 3A through 8A, make up the p-block of elements. In these elements, s orbitals are filled, and the highest-energy electrons are in p orbitals.

The transition metals are the d-block elements. In these elements, the highest-energy electrons are in the d sublevel of the energy level one less than the period number. Most d-block elements have two electrons in s orbitals, but in some, such as chromium and copper, the d sublevel "borrows" an electron from the s orbital to form half-filled (Cr) or filled (Cu) d orbitals. The remaining block is the f-block, or inner transition metals. The highest-energy electrons in these elements are in an f sublevel of the energy level two less than the period number.

The following Example Problem illustrates how electron configuration determines an element's position in the periodic table.

Example Problem 6-1
Electron Configuration and the Periodic Table

The electron configuration of phosphorus is $[Ne]3s^23p^3$. Without using the periodic table, determine the group, period, and block in which the element is located in the periodic table.

First, identify the valence electrons and note their energy level. In phosphorus, the 3 in front of the s and p orbitals indicates that the valence electrons are in the third energy level. Therefore, phosphorus will be found in the third period of the periodic table.

Next, note the sublevel of the highest-energy electrons. In the case of phosphorus, these electrons are in a p sublevel. Therefore, phosphorus will be found in the p-block.

Finally, use the number of valence electrons to determine the group number of the element. Phosphorus has two electrons in an s orbital and three electrons in p orbitals for a total of five valence electrons. Because there are no incomplete d or f sublevels, phosphorus must be a representative element in group 5A.

To summarize, phosphorus is found in period 3, group 5A, and the p-block of the periodic table. A glance at the table will confirm this answer.

Practice Problems

11. Without using the periodic table, determine the group, period, and block in which an element with each of the following electron configurations is found.

 a. $[He]2s^2 2p^5$ **c.** $[Kr]5s^2 4d^{10} 5p^3$

 b. $[Ar]4s^2$ **d.** $[Ar]4s^2 3d^3$

12. Write the electron configuration of the following elements.

 a. the alkaline earth element in the sixth period

 b. the halogen in the third period

 c. the group 4A element in the third period

 d. the group 5B element in the fourth period

 e. the group 1A element in the fifth period

6.3 Periodic Trends

The electron structure of an atom determines many of its chemical and physical properties. Because the periodic table reflects the electron configurations of the elements, the table also reveals trends in the elements' chemical and physical properties.

▶**Atomic radius** The atomic radius is a measure of the size of an atom. The larger the radius, the larger is the atom. Research shows that atoms tend to decrease in size across a period because the nuclei are increasing in positive charge while electrons are being added to sublevels that are very close in energy. As a result, the increased nuclear charge pulls the outermost electrons closer to the nucleus, making the atom smaller.

Moving down through a group, atomic radii increase. Even though the positive charge of the nucleus increases, each successive element has electrons in the next higher energy level. Electrons in these higher energy levels are located farther from the nucleus than those in lower energy levels. The increased size of higher energy level outweighs the increased nuclear charge. Therefore, the atoms increase in size.

Practice Problems

13. For each of the following pairs, predict which atom is larger.

a. Mg, Sr   **d.** Ge, Br

b. Sr, Sn   **e.** Cr, W

c. Ge, Sn

14. Comparing elements from left to right across a period, what general trend would you predict for the energy required to remove a valence electron from an atom? Explain the basis for your prediction.

▶ **Ionic radius** When an atom gains or loses one or more electrons, it becomes an ion. Because an electron has a negative charge, gaining electrons produces a negatively charged ion, whereas losing electrons produces a positively charged ion. As you might expect, the loss of electrons produces a positive ion with a radius that is smaller than that of the parent atom. Conversely, when an atom gains electrons, the resulting negative ion is larger than the parent atom.

Practically all of the elements to the left of group 4A of the periodic table commonly form positive ions. As with neutral atoms, positive ions become smaller moving across a period and become larger moving down through a group. Most elements to the right of group 4A (with the exception of the noble gases in group 8A) form negative ions. These ions, although considerably larger than the positive ions to the left, also decrease in size moving across a period. Like the positive ions, the negative ions increase in size moving down through a group.

Practice Problems

In the following questions, the charges of ions are indicated by the superscript numbers and signs.

15. For each of the following pairs, predict which atom or ion is larger.

a. Mg, Mg^{2+} **d.** Cl^-, I^-

b. S, S^{2-} **e.** Na^+, Al^{3+}

c. Ca^{2+}, Ba^{2+}

16. Predict which of the ions, Mg^{2+} or S^{2-}, is larger. Explain your prediction.

▶ **Ionization energy** Energy is required to pull an electron away from an atom. The first ionization energy of an element is the amount of energy required to pull the first valence electron away from an atom of the element. Atoms with high ionization energies, such as fluorine, oxygen, and chlorine, are found on the right side of the periodic table and are unlikely to form positive ions by losing electrons. Instead, they usually gain electrons, forming negative ions.

Atoms with low ionization energies, such as sodium, potassium, and strontium, lose electrons easily to form positive ions and are on the left side of the periodic table. Recall that atoms decrease in size from left to right across a period. First ionization energies generally increase across a period of elements primarily because the electrons to be removed are successively closer to the nucleus. First ionization energies decrease moving down through a group of elements because the sizes of the atoms increase and the electrons to be removed are farther from the nucleus.

Practice Problems

17. For each of the following pairs, predict which atom has the higher first ionization energy.

    a. Mg, Na   **d.** Cl, I

    b. S, O   **e.** Na, Al

    c. Ca, Ba   **f.** Se, Br

18. For each of the following pairs, predict which atom forms a positive ion more easily.

    a. Be, Ca   **d.** K, Ca

    b. F, I   **e.** Sr, Sb

    c. Na, Si   **f.** N, As

▶ **The octet rule** When atoms lose or gain electrons, they generally do so until the ion has eight valence electrons—the stable s^2p^6 electron configuration of a noble gas. This principle is called the **octet rule.** Exceptions to this rule are hydrogen, which can gain an electron, obtaining the stable $1s^2$ configuration of helium, and elements in period 2, such as lithium and beryllium, that lose electrons, also obtaining the helium configuration.

The octet rule lets you predict the ionic charge of a representative element. For example, you can predict that an element in group 6A, having a high ionization energy, will gain two electrons to achieve a stable octet configuration.

Practice Problems

19. For each of the following elements, state whether it is more likely to gain or lose electrons to form a stable octet configuration and how many electrons will be gained or lost.

a. K
b. Br
c. O
d. Mg

e. Al
f. I
g. Ar

20. Which noble-gas configuration is each of the following elements most likely to attain by gaining or losing electrons?

a. S
b. Sr
c. Cl
d. Be

e. Fr
f. N
g. Ba

▶ **Electronegativity** When atoms combine chemically with each other, they do so by forming a chemical bond. This bond involves either the transfer of electrons or sharing of electrons to varying degrees. The nature of the bond between two atoms depends on the relative ability of each atom to attract electrons from the other, a property known as **electronegativity.**

The maximum electronegativity value is 3.98 for fluorine, the element that attracts electrons most strongly in a chemical bond. The trends in electronegativity in the periodic table are generally similar to the trends in ionization energy. The lowest electronegativity values occur among the elements in the lower left of the periodic table. These atoms, such as cesium, rubidium, and barium, are large and have few valence electrons, which they lose easily. Therefore, they have little attraction for electrons when forming a bond.

Elements with the highest electronegativity values, such as fluorine, chlorine, and oxygen, are found in the upper right of the periodic table (excluding, of course, the noble gases, which do not normally form chemical bonds). These atoms are small and can gain only one or two electrons to have a stable noble-gas configuration. Therefore, when these elements form a chemical bond, their attraction for electrons is large.

Electronegativities generally increase across a period and decrease down through a group.

Practice Problems

21. For each of the following pairs, predict which atom has the higher electronegativity.

   a. Mg, Na   **d.** Ca, Ba
   b. Na, Al   **e.** S, O
   c. Cl, I   **f.** Se, Br

Chapter 6 Review

22. Explain why the word *periodic* is applied to the table of elements.

23. Why do elements in a group in the periodic table exhibit similar chemical properties?

24. What chemical property is common to the elements in group 8A of the periodic table? Why do these elements have this property?

25. In terms of electron configurations, what does the group number of the A-groups in the periodic table tell you?

26. Describe the group and period trends in the following atomic properties.

   a. atomic radius
   b. electronegativity
   c. first ionization energy
   d. ionic radius

27. Describe the relationship between the electronegativity value of an element and the tendency of that element to gain or lose electrons when forming a chemical bond.

The Elements

7.1 Properties of s-Block Elements

There are 92 naturally occurring elements in the universe. The remaining synthetic elements are created in laboratories or nuclear reactors. The elements in groups 1A through 8A of the periodic table display a wide range of properties and are called representative elements. The number of valence electrons in these elements ranges from one in group 1A to eight in group 8A, corresponding to the group numbers. The electrons are in s or p orbitals.

Because elements in a given group have the same number of valence electrons, they have similar properties. The properties are not identical, however, because the numbers of nonvalence electrons differ. For example, the ionization energy of elements in a group decreases as the atomic number increases. For that reason, metals, which tend to lose electrons when they react, increase in reactivity as the atomic number increases. Nonmetals, which tend to gain electrons, decrease in reactivity as the atomic number increases.

Sometimes a period-2 element has more properties in common with a period-3 element in a neighboring group than with the period-3 element in its own group. Such a relationship is called a **diagonal relationship.**

▶ **Hydrogen** Although it is usually placed in group 1A, the element hydrogen is not a member of any group because it has properties of both metals and nonmetals. Hydrogen behaves as a metal when it loses its single electron. It behaves as a nonmetal when it gains an electron. The universe contains more than 90 percent hydrogen by mass. Hydrogen can react explosively with oxygen. The product of this reaction is water. The main industrial use of hydrogen is in the production of ammonia.

▶ **Group 1A: Alkali metals** The alkali metals are soft metals that are highly reactive, easily losing a single valence electron to form 1+ ions. Lithium, the least reactive alkali metal, has a diagonal relationship with magnesium, a group 2A metal. Sodium and potassium are the most abundant alkali metals. The most common sodium compound is sodium chloride, or table salt. Potassium compounds are included in fertilizers because potassium is essential for plant growth. The remaining alkali metals—rubidium, cesium, and francium—have relatively few commercial uses.

▶ **Group 2A: Alkaline earth metals** The group 2A elements are hard, shiny solids called alkaline earth metals. Although not as reactive as alkali metals, they easily lose two valence electrons to form 2+ ions. Beryllium is used to moderate neutrons in nuclear reactors and in alloys used for non-sparking tools. Calcium is an essential element for humans, especially in maintaining healthy bones and teeth. Calcium carbonate is found in rocks such as limestone. It is used in antacids and as an abrasive. Lime, an oxide of calcium, is used to make soil less acidic, to remove pollutants from smokestacks, and to make mortar. Magnesium is used in lightweight alloys and forms an oxide that is extremely heat-resistant.

Example Problem 7-1
Comparing the s-Block Elements

Describe the following pair of elements in terms of group number, number of valence electrons, typical ion formed, ionization energy, and reactivity.

Sodium (Na) and potassium (K)

Both sodium and potassium are in group 1A. They each have one valence electron and form 1+ ions. Sodium is above potassium in the group; thus, sodium has a higher ionization energy and is less reactive than potassium.

Practice Problems

1. Compare each of the following pairs of elements in terms of group number, number of valence electrons, typical ion formed, ionization energy, and reactivity.

   a. calcium (Ca) and strontium (Sr)

   b. lithium (Li) and francium (Fr)

   c. cesium (Cs) and barium (Ba)

Example Problem 7-2
Using Properties to Identify s-Block Elements

Two elements, X and Y, have the following properties:
X is a gas at room temperature and forms 1+ ions and 1− ions. Y is a solid at room temperature, has many properties in common with magnesium, forms only 1+ ions, and is the least reactive element in its group. Identify the elements.

X must be hydrogen because of all s-block elements, only hydrogen can form 1− ions and is a gas at room temperature. Y also is in group 1A because it forms 1+ ions. The similarity of Y to magnesium suggests a diagonal relationship. The additional fact that Y is the least reactive element in its group make it clear that Y is lithium.

Practice Problems
2. Identify the following group 2A elements based on the properties described.
    a. an element that forms an oxide used to line furnaces
    b. an element with two valence electrons that is used to moderate neutrons in nuclear reactors

7.2 Properties of p-Block Elements

The elements of the p-block are in groups 3A through 8A and include metals, nonmetals, and metalloids. Many can form more than one type of ion.

▶ **Group 3A: The boron group** The group 3A elements have three valence electrons and include the metalloid boron and the metals aluminum, gallium, indium, and thallium. Boron and aluminum form 3+ ions, gallium and indium form both 3+ and 1+ ions, and thallium forms only 1+ ions.

The main source of boron is a compound called borax, which is used as a cleaning agent and as fireproof insulation. Aluminum, the most abundant metal, is used to make products such as cans. Gallium is used in some thermometers because it remains liquid over a wide temperature range.

▶ **Group 4A: The carbon group** The group 4A elements have four valence electrons. Carbon is one of the most important nonmetallic elements. A branch of chemistry called organic chemistry studies the carbon compounds that control what happens in cells. Elemental carbon occurs in various forms including the soft graphite used in pencils, and diamond, one of the hardest substances known. Forms of an element in the same physical state that have different structures and properties are called **allotropes.**

Group 4A also contains the metalloids silicon and germanium and the metals tin and lead. Silicon, the second most abundant element in Earth's crust, is most often combined with oxygen in silica, which is found in quartz, sand, and glass. Alloys of tin such as bronze are mainly used for decorative items. Lead was one of the first metals obtained from its ore and had many uses until it was determined to be toxic. The major current use of lead is in automobile storage batteries.

▶ **Group 5A: The nitrogen group** The elements in group 5A have five valence electrons and include the nonmetals nitrogen and phosphorus, the metalloids arsenic and antimony, and the metal bismuth. Nitrogen gas makes up most of Earth's atmosphere. Bacteria in soil and roots convert molecular nitrogen into compounds that can be used by plants and the animals that consume the plants. Nitrogen is used to make ammonia and nitric acid, which is used to produce fertilizers, dyes, and explosives. Compounds of phosphorus can be found in baking powder, cleaning products, and fertilizer.

Example Problem 7-3
Identifying Elements in Groups 3A, 4A, and 5A

Find the following pairs of elements in the periodic table. State the name of each. Then compare them in terms of group number and number of valence electrons. Identify each element as a metal, nonmetal, or metalloid. Also state a use for each element.

a. C and Pb   **b.** Si and P   **c.** Ga and N

a. Both C, carbon, and Pb, lead, are in group 4A and have four valence electrons. Carbon is a nonmetal. In its graphite form, it is used in pencils. Lead is a toxic metal used in automobile storage batteries.

b. Si, silicon, is in group 4A and has four valence electrons. P, phosphorus, is in group 5A and has five valence electrons.

Silicon is a metalloid used in computer chips. Phosphorus is a nonmetal. Red phosphorus provides the striking surface for matchboxes.

c. Ga, gallium, is in group 3A and has three valence electrons. N, nitrogen, is in group 5A and has five valence electrons. Gallium is a metal used in some thermometers; a compound of gallium is used to produce semiconductor chips used in light-powered calculators. Nitrogen is a nonmetal used to make ammonia, which is found in many cleaning products.

Practice Problems

3. Name the elements in each of the following pairs. Compare them in terms of group number, number of valence electrons, and metallic character.

   a. As and Bi   **b.** Ge and N   **c.** B and Sn

4. An element has three valence electrons. It always loses one electron per atom when it forms an ion, and it behaves like a metal. Identify the element's group number, and name the element.

▶ **Group 6A: The oxygen group** The elements in group 6A have six valence electrons. These elements include the nonmetals oxygen and sulfur, the metalloids selenium and tellurium, and the rare radioactive metal polonium. Oxygen is the most abundant element in Earth's crust. It has two allotropes. One, O_2, supports combustion and is used by organisms to release energy during respiration. The other allotrope, ozone, is an unstable, irritating gas. Sulfur, which has ten allotropes, is used to make sulfuric acid, a compound whose production is an indication of the strength of an economy. Selenium is used in solar panels and photocopiers.

▶ **Group 7A: The halogens** The group 7A elements have seven valence electrons and include the highly reactive nonmetals fluorine, chlorine, bromine, and iodine and the rare radioactive element astatine. Compounds of fluorine, the most electronegative element, are used in toothpaste. Chlorine is used as a disinfectant and bleach and to make certain plastics. The silver compounds of bromine and iodine are used to coat photographic film. Iodine is important for maintaining a healthy thyroid gland.

▶ **Group 8A: The noble gases** The noble gases have filled outer energy levels and are extremely unreactive. Helium is used in balloons and in breathing mixtures for divers. Neon and other noble gases are used to produce colored light displays. Argon, the most abundant noble gas, provides an inert atmosphere when a mixture of oxygen, heat, and sparks would be dangerous. Argon and krypton are used to extend the life of filaments in incandescent lightbulbs.

Example Problem 7-4
Identifying Elements in Groups 6A, 7A, and 8A

Find the following pairs of elements in the periodic table. Give the name of each element. Then compare them in terms of group number and number of valence electrons. State what negatively charged ion, if any, each element forms. State a use for each element.

a. O and Ne   **b.** Br and Kr

a. O, oxygen, is in group 6A, has six valence electrons, and tends to form 2− ions. Oxygen is used by organisms to release energy during respiration. Ne, neon, is in group 8A, has eight valence electrons, and does not tend to form ions. It is used in colored light displays.

b. Br, bromine, is in group 7A, has seven valence electrons, and tends to form 1− ions. A compound of bromine and silver is used to coat photographic film. Kr, krypton, is in group 8A, has eight valence electrons, and does not tend to form ions. Krypton is used to prolong the life of filaments in incandescent lightbulbs.

Practice Problems
5. Name the elements in each of the following pairs. Compare them in terms of group number, number of valence electrons, and typical negative ion formed if any. Also, state a use for each element.

    a. Se and Cl   **b.** I and H   **c.** S and F

6. An element is a gas at room temperature. It does not form any compounds. It has eight valence electrons and is higher in atomic mass than the element phosphorus but lower in mass than arsenic. Identify the element's group name and number, and name the element.

7. An element is metallic and radioactive. It has six valence electrons. Identify the element's group number and name the element.

7.3 Properties of d-Block and f-Block Elements

The transition elements make up the B groups of the periodic table. They are subdivided into d-block and f-block elements—the transition metals and inner transition metals, respectively.

▶ **Transition metals** The final electrons of transition metals enter a d sublevel. The transition metals have typical metallic properties, such as malleability and electrical conductivity. Most are hard and have high melting and boiling points.

Transition metals can lose their two s electrons and form 2+ ions. Some can also lose d electrons and take on higher charges. The positive ions that have unpaired d electrons are typically colored.

The movement of electrons in metals can give rise to magnetism. If all electrons are paired, the metal is diamagnetic; that is, it is not attracted to or is slightly repelled by a magnetic field. If there is an unpaired electron, the metal is paramagnetic and is slightly attracted to a magnetic field. A few metals, such as iron and cobalt, exhibit **ferromagnetism,** a powerful attraction to a magnetic field.

▶ **Inner transition metals** The final electrons of the inner transition metals enter an f sublevel. The inner transition metals are placed below the main body of the periodic table. They include the **lanthanide series,** which are in period 6 and follow the element lanthanum, and the **actinide series,** which are in period 7 and follow actinium. The lanthanides have very similar properties and are silvery metals with relatively high melting points. They are used in welder's glasses, television screens, and lasers. The actinides are all radioactive, and most are synthetic elements. They include the **transuranium elements,** which are elements that have an atomic number greater than 92.

Example Problem 7-5
Characteristics of Transition Metals
Describe the following transition metal in terms of period, block, number and pairing of d electrons, magnetic properties: scandium (Sc, atomic number 21).

Scandium is in period 4 and is a d-block element. Scandium tends to be paramagnetic and has only one d electron, which must be unpaired.

Practice Problems

8. Compare the following pair of transition metals in terms of period number, block, number and pairing of d electrons, and magnetic properties.

Nickel (Ni, atomic number 28) and yttrium (Y, atomic number 39)

9. For each of the following transition elements, state the period number and block, and tell whether each is a transition metal, a lanthanide, or an actinide.

a. samarium (Sm, atomic number 62)

b. fermium (Fm, atomic number 100)

c. osmium (Os, atomic number 76)

Chapter 7 Review

10. Explain why elements within the same group of the periodic table are similar but not identical in properties, such as ionization energy.

11. State what is meant by a diagonal relationship and give an example of one.

12. Compare the alkali and alkaline earth metals in terms of position in the periodic table, number of valence electrons, and overall properties.

13. Compare the metallic character of the elements carbon, silicon, and lead. What do these elements have in common in terms of valence electrons and placement in the periodic table?

14. What is meant by the term *allotropes*? Give an example of allotropes.

15. Compare the period-2 elements that are in groups 5A, 6A, 7A, and 8A in terms of number of valence electrons and reactivity.

16. What is the difference between the transition metals and inner transition metals in terms of their final electron? Describe the placement of transition metals and inner transition metals in the periodic table and give an example of each.

Ionic Compounds

8.1 Forming Chemical Bonds

The force that holds two atoms together is called a **chemical bond.**
Chemical bonds form because of attractions between oppositely
charged atoms, called ions, or between electrons and nuclei. The
outermost, or valence, electrons of atoms are the ones mainly
involved in the formation of bonds. The elements within a group
of the periodic table typically have the same number of valence
electrons.

Elements tend to react so as to achieve the stable electron configuration of a noble gas, typically an octet of electrons. A **cation,** or
positive ion, is formed when an atom loses one or more electrons.
An **anion,** or negative ion, is formed when an atom gains one or
more electrons. The periodic table is useful in predicting the charges
of ions typically formed by various atoms.

Example Problem 8-1
Determining Charges of Ions

Calcium (Ca, atomic number 20) is an element in group 2A of the
periodic table. Write the electron configuration for a neutral atom of
calcium. Tell how many electrons this atom readily tends to gain or
lose to form an ion. Predict the charge on the ion, write its formula,
and tell whether it is a cation or an anion. Finally, write the electron
configuration of this ion.

A neutral atom of element 20 would have 20 electrons, giving
it the electron configuration $1s^22s^22p^63s^23p^64s^2$. To achieve the
stable electron configuration of a noble gas, the atom would tend
to lose its two valence electrons, producing a cation with a charge
of $2+$ and the formula Ca^{2+}. The configuration of this ion would
be $1s^22s^22p^63s^23p^6$.

Practice Problems

For each of the following atoms, write the electron configuration, referring to the periodic table. Then write the formula of the ion the atom is most likely to form and identify that ion as a cation or an anion. Finally, write the electron configuration of the ion.

1. bromine (Br), element 35

2. gallium (Ga), element 31

3. sulfur (S), element 16

4. rubidium (Rb), element 37

8.2 The Formation and Nature of Ionic Bonds

To bond ionically, atoms must transfer valence electrons. An atom that loses one or more electrons becomes a positive ion. An atom that gains one or more electrons becomes a negative ion. The **ionic bond** that forms is the electrostatic force holding the oppositely charged ions together.

The total number of electrons lost must equal the total number of electrons gained. The ratio of atoms that bond ionically must therefore be such that overall electrical neutrality is maintained.

Example Problem 8-2
The Formation of an Ionic Compound

Atoms of magnesium (Mg, atomic number 12, group 2A) and chlorine (Cl, atomic number 17, group 7A) bond to form the ionic compound magnesium chloride. Use electron configurations and the balance of charges to determine the ratio of magnesium and chlorine atoms in magnesium chloride.

A neutral atom of magnesium has the electron configuration $1s^2 2s^2 2p^6 3s^2$, or, in abbreviated form, $[Ne]3s^2$. Chlorine has the electron configuration $[Ne]3s^2 3p^5$. An atom of Mg would lose its two valence electrons to form a $2+$ ion and achieve the stable electron configuration of a noble gas. An atom of Cl would gain one electron to form a $1-$ ion. To achieve overall neutrality in the compound, each Mg ion would require two Cl ions: $(1 \times 2+) + (2 \times 1-) = 0$. Thus, there is one Mg atom for every two Cl atoms in magnesium chloride.

Practice Problems

Write the electron configurations, in abbreviated form, for the atoms in each pair, referring to the periodic table as necessary. Then determine the ratio of the atoms in the ionic compound formed in each case.

5. aluminum (Al) and fluorine (F)

6. lithium (Li) and oxygen (O)

7. beryllium (Be) and selenium (Se)

8. gallium (Ga) and sulfur (S)

▶ **Properties of ionic compounds and lattice energy** In a solid ionic compound, the positive ions are surrounded by negative ions, and the negative ions by positive ions. The resulting structure is called a crystal lattice and contains a regular, repeating, three-dimensional arrangement of ions. This arrangement, which involves strong attraction between oppositely charged ions, tends almost always to produce certain properties, such as high melting and boiling points and brittleness. Ionic compounds are always non-conductors of electricity when solid but good conductors when melted. They also act as electrolytes, substances that conduct electric current when dissolved in water. The combination of these conductivity characteristics is a very good identifier of ionic compounds, although each characteristic separately is not very reliable.

The energy required to separate one mole of the ions of an ionic compound is called **lattice energy,** which is expressed as a negative quantity. The greater (that is, the more negative) the lattice energy is, the stronger is the force of attraction between the ions. Lattice energy tends to be greater for more-highly-charged ions and for small ions than for ions of lower charge or large size.

Practice Problems

9. On the basis of the properties of the following "unknowns," classify each as either ionic or not ionic.

  a. conducts electricity when solid

  b. conducts electricity when liquid and has a low melting point

  c. has a high boiling point and shatters when hammered

  d. has a high melting point and conducts electricity when dissolved in water

10. For each of the following pairs of ionic compounds, state which would be expected to have the higher (more negative) lattice energy.

a. LiF or KBr

b. NaCl or MgS

c. MgO or RbI

8.3 Names and Formulas for Ionic Compounds

The simplest ratio of the ions represented in an ionic compound is called a **formula unit.** The overall charge of any formula unit is zero. In order to write a correct formula unit, one must know the charge of each ion. The charges of **monatomic ions,** or ions containing only one atom, can often be determined by referring to the periodic table or table of common ions based on group number. For example, ions of group 1A typically have a charge of 1+. Those of group 2A have a charge of 2+. Those of group 7A have a charge of 1−.

The charge of a monatomic ion is equal to its **oxidation number.** The oxidation number, or oxidation state, of an ion in an ionic compound is numerically equal to the number of electrons that were transferred to or from an atom of the element in forming the compound. If the electrons were transferred from the atom, the ion has a positive oxidation state. If they were transferred to the atom, the ion has a negative oxidation state. Most transition metals and metals of groups 3A and 4A can have more than one oxidation number.

Oxidation numbers can be used to determine the chemical formulas for ionic compounds. If the oxidation number of each ion is multiplied by the number of that ion present in a formula unit, and then the results are added, the sum must be zero.

In the formula for an ionic compound, the symbol of the cation is written before that of the anion. Subscripts, or small numbers written to the lower right of the chemical symbols, show the numbers of ions of each type present in a formula unit.

Copyright © Glencoe/McGraw-Hill, a division of the McGraw-Hill Companies, Inc.

Example Problem 8-3
Determining the Formula for an Ionic Compound

Determine the correct chemical formula for the ionic compound formed from strontium (Sr) and bromine (Br).

Strontium is an element in group 2A of the periodic table. Bromine is in group 7A. The formulas for their ions are therefore Sr^{2+} and Br^-, respectively. Sr must lose two electrons to form its ion, and Br must gain one electron. The total number of electrons lost must equal the total number of electrons gained so that overall charge is zero. Subscripts must be chosen so that this is the case. For electrical neutrality, there must be two Br for each Sr. The correct formula is thus $SrBr_2$, where the subscript 1 is understood after the Sr. This formula can be verified by multiplying the subscripts by the ion charges and summing the result, as follows.

$$(1 \times 2+) + (2 \times 1-) = 0$$

Practice Problems

Write the correct formula for the ionic compound formed between atoms of each of the following pairs of elements.

11. sodium (Na) and sulfur (S)

12. magnesium (Mg) and nitrogen (N)

13. potassium (K) and phosphorus (P)

14. barium (Ba) and fluorine (F)

15. aluminum (Al) and nitrogen (N)

▶ **Polyatomic ions** An ion that contains more than one atom is called a **polyatomic ion.** The charge on such an ion applies to the entire group of atoms. The writing of chemical formulas for compounds containing such ions follows the same rules as for compounds containing only monatomic ions. The overall charge must be zero. Subscripts within the formula for a polyatomic ion must not be changed during formula writing. If there is more than one such ion in a formula unit, parentheses are written around the formula of the ion and a subscript is written after the final parenthesis.

Example Problem 8-4
The Formula for a Compound Containing a Polyatomic Ion

Write the chemical formula for a compound formed from ammonium ions and sulfate ions. Refer to a table of common ions as necessary.

The ammonium ion has the formula NH_4^+. The sulfate ion has the formula SO_4^{2-}. The overall charge of the formula unit must be zero, which means that there must be two ammonium ions for each sulfate ion. Thus, the correct formula is $(NH_4)_2SO_4$. The subscript after the parenthesis indicates the presence of two ammonium ions.

Practice Problems
Write the correct formula for the ionic compounds that contain the following ions. Refer to a table of polyatomic ions as necessary.

16. sodium and phosphate

17. ammonium and carbonate

18. aluminum and chromate

19. calcium and hydroxide

▶ **Naming ionic compounds** In naming ionic compounds, name the cation first, then the anion. Monatomic cations use the element name. Monatomic anions use the root of the element name plus the suffix *-ide*. If an element can have more than one oxidation number, use a Roman numeral in parentheses after the element name, for example, iron(II) to indicate the Fe^{2+} ion. For polyatomic ions, use the name of the ion.

Certain polyatomic ions, called **oxyanions**, contain oxygen and another element. If two different oxyanions can be formed by an element, the suffix *-ate* is used for the oxyanion containing more oxygen atoms, and the suffix *-ite* for the oxyanion containing fewer oxygens. In the case of the oxyanions of the halogens, the following special rules are used.

four oxygens, *per* + root + *-ate* (example: perchlorate, ClO_4^-)

three oxygens, root + *-ate* (example: chlorate, ClO_3^-)

two oxygens, root + *-ite* (example: chlorite, ClO_2^-)

one oxygen, *hypo-* + root + *-ite* (example: hypochlorite, ClO^-)

Practice Problems

Name the ionic compounds that have the following formulas.

20. NH_4I

21. $NaBrO_3$

22. $Mg(NO_3)_2$

23. $KHSO_4$

24. NH_4ClO_4

25. $Al(ClO)_3$

26. FeF_2

8.4 Metallic Bonds and Properties of Metals

The bonding in metals is explained by the **electron sea model,** which proposes that the atoms in a metallic solid contribute their valence electrons to form a "sea" of electrons that surrounds metallic cations. These **delocalized electrons** are not held by any specific atom and can move easily throughout the solid. A **metallic bond** is the attraction between these electrons and a metallic cation.

Many of the properties of metals can be explained by means of metallic bonds and delocalized electrons. For example, metals generally have extremely high boiling points because it is difficult to pull metal atoms completely away from the group of cations and attracting electrons. The melting points are considerably lower because of the mobility of the cations and electrons, which can move past each other. Metals are also malleable (able to be hammered into sheets) and ductile (able to be drawn into wire) because of the mobility of the particles. The delocalized electrons make metals good conductors of electricity. These electrons also interact with light, which is why metals tend to be highly lustrous.

A mixture of elements that has metallic properties is called an **alloy.** Alloys can be of two basic types. A substitutional alloy is one in which atoms of the original metal are replaced by other atoms of similar size. An interstitial alloy is one in which the small holes in a metallic crystal are filled by other smaller atoms.

Practice Problems

27. An unknown substance is found to be a good conductor of electricity when melted. Explain whether on this basis alone the substance can be classified as a metal.

28. Suppose the substance in question 27 is allowed to solidify. It is then hammered, and shatters as a result. What property is being tested? Would you classify this material as a metal? Explain your answer.

29. A certain sample of an element is known to be either nickel, sulfur, or iodine. The sample is found to be a good conductor of electricity as a solid. Which element is it likely to be? Why? What other properties would you expect the sample to have?

30. Carbon steel is made by filling the holes in an iron crystal with atoms of carbon. As what type of material is such a mixture classified?

31. Brass is a mixture in which some of the atoms in a crystal of copper are replaced by zinc atoms, which are of similar size. As what type of material is such a mixture classified?

Chapter 8 Review

32. Define the term *chemical bond*. Account for such bonding in terms of the concept of noble gas configuration.

33. Explain the difference between a cation and an anion in terms of electron transfer. Give an example of each type of ion.

34. Define the term *ionic bond*. Explain how atoms of potassium (group 1A) and iodine (group 7A) bond ionically with one another.

35. Describe the arrangement of ions in ionic compounds and the properties typical of ionic compounds.

36. Explain how to determine the formula for an ionic compound, given its name.

37. State the rules for naming oxyanions, including those of the halogens.

38. Describe the model for metallic bonding and explain how the model accounts for typical metallic properties.

39. Compare and contrast substitutional and interstitial alloys.

Covalent Bonding

9.1 The Covalent Bond

Atoms typically bond to achieve the stable configuration of a noble gas, generally an octet of electrons. Two atoms that both need to gain valence electrons may share them in a bond called a **covalent bond.** When two or more atoms bond covalently, a **molecule** is formed. In the formation of a covalent bond, the atoms move close enough so that the repulsive forces due to like-charged particles are balanced by the attractive forces between oppositely charged particles.

Pairs of electrons involved in forming covalent bonds are called bonding pairs. When a single pair of electrons is shared, a single covalent bond results. Remaining pairs of electrons not involved in bonding are called lone pairs. The arrangement of electrons in molecules can be illustrated by **Lewis structures,** which use electron-dot diagrams. In such diagrams, shared electron pairs are represented as a pair of dots or a line.

Example Problem 9-1
Drawing a Lewis Structure for a Molecule

What is the Lewis structure for a molecule of the covalently bonded compound nitrogen triiodide (NI_3), given that all the bonds are single? State how many single covalent bonds there are and how many lone pairs each bonded atom has.

As a group 5A element, nitrogen has five valence electrons and therefore requires three more electrons to achieve a complete octet. It must form three covalent bonds to do so. Iodine, a group 7A element, has seven valence electrons and needs one more electron to achieve a complete octet. It must form only one bond to do so. The Lewis structure for NI_3 is shown below.

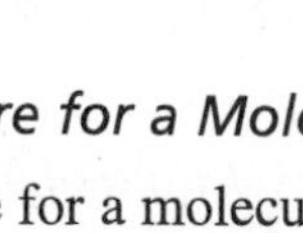

$$:\ddot{I}—\ddot{N}—\ddot{I}:$$
$$|$$
$$:\ddot{I}:$$

There are three single covalent bonds in the molecule. There is one lone pair on the nitrogen atom and three lone pairs on each iodine atom.

Practice Problems

1. Draw the Lewis structure for each of the following molecules, which contain only single covalent bonds. Refer to a periodic table as necessary. Also state how many single covalent bonds there are and how many lone pairs are on each bonded atom.
   a. HCl
   b. SCl_2
   c. PH_3
   d. SiF_4

▶ **Bond types and multiple bonds** Single bonds are always **sigma bonds,** bonds in which the electron pair is shared in an area centered between the two atoms. Such bonds can form when the bonding orbital is created by overlap of two s orbitals, an s and a p orbital, or two p orbitals.

   In some cases, atoms attain a noble gas configuration by sharing more than one pair of electrons, forming a multiple covalent bond. In a double covalent bond, two pairs of electrons are shared. In a triple covalent bond, three pairs are shared. A multiple covalent bond always consists of a sigma bond and at least one **pi bond,** a bond in which parallel orbitals overlap. A pi bond occupies space above and below the line that represents where the two atoms are joined.

Example Problem 9-2
Identifying Bond Types in a Molecule

The oxygen molecule has the formula O_2. In this molecule, a total of four electrons are shared by the two atoms. What type of bond is this: single, double, or triple? Is the bond made up of sigma bonds, pi bonds, or both?

The four shared electrons make up two bonding pairs, which means that the bond is a double bond. In any multiple bond, there is one sigma bond and the remaining bonds are pi bonds. Thus, in this case, there are one sigma bond and one pi bond.

Practice Problems

2. Identify the bond types (single, double, or triple; and sigma or pi) in each of the following diatomic molecules, given the number of shared electrons.

   a. N_2, six shared electrons

   b. Br_2, two shared electrons

Copyright © Glencoe/McGraw-Hill, a division of the McGraw-Hill Companies, Inc.

c. HCN, in which C is the central atom, two electrons are shared by H and C, and six electrons are shared by C and N

d. CO_2, in which C is the central atom, four electrons are shared by C and the first O, and four electrons are shared by C and the second O

▶ **Bond length and strength** The distance between the nuclei of two bonded atoms is called bond length. When the balance of forces between the atoms is sufficiently upset, the bond can break. Energy is absorbed when a bond breaks and is released when a bond forms. The amount of energy required to break a covalent bond is called the bond dissociation energy and always has a positive value. The stronger the bond, the greater is the bond dissociation energy, and, therefore, the more difficult it is to break the bond. For similar-sized atoms, it is typically the case that a shorter bond has a greater bond dissociation energy than does a longer bond. Triple bonds are usually shorter and have greater bond dissociation energies than do double bonds. Double bonds are usually shorter and have greater bond dissociation energies than do single bonds.

The total energy change in a chemical reaction is determined from the energy of the bonds broken and formed. In an **endothermic reaction,** more energy is required to break existing bonds in the reactants than is released when the new product bonds form. In an **exothermic reaction,** more energy is released in the formation of the new product bonds than is required to break bonds in the reactants.

Practice Problems

3. The bond length between two double-bonded nitrogen atoms within a molecule of a certain compound is less than that between two single-bonded carbon atoms within another molecule. Given that N and C are next to each other in the periodic table, how would you expect the bond dissociation energies and bond strengths of these two bonds to compare?

4. A molecule of ethene (C_2H_4) contains a double bond between the two carbon atoms. A molecule of ethyne (C_2H_4) contains a triple bond between the two carbon atoms. How would you expect the lengths, strengths, and dissociation energies of these two bonds to compare?

5. In a certain reaction, 100 kJ is required to break the reactant bonds, and 130 kJ is released when the new product bonds form. Is the reaction exothermic or endothermic?

9.2 Naming Molecules

Binary molecular compounds are those that contain only two different elements. In naming such compounds, given their formulas, the rules below are followed.

1. Name the first element in the formula first, using its element name, unchanged.
2. Name the second element, using the root of its name and adding the suffix *-ide*.
3. Use the following prefixes, given the number of each type of atom present: 1, *mono-*; 2, *di-*; 3, *tri-*; 4, *tetra-*; 5, *penta-*; 6, *hexa-*; 7, *hepta-*; 8, *octa-*; 9, *nona-*; and 10, *deca-*. *Mono-* is not used as a prefix for the name of the first element.

Example Problem 9-3
Naming Binary Molecular Compounds

Name the binary molecular compound that has the formula S_2O_7.

The first element is named first, using the unchanged element name: *sulfur*. The second element's root is then written with the suffix *-ide*: *ox-* plus *-ide* = oxide. There are two sulfur atoms, so the prefix *di-* is added. There are seven oxygen atoms, so the prefix *hepta-* is added. Thus, the complete name is disulfur heptoxide. (The *a* in *hepta-* is dropped before the vowel *o* in *oxide*.)

Practice Problems

6. Name each of the following binary molecular compounds.

   a. SO_2 **b.** P_4O_{10} **c.** N_2O_3 **d.** SiF_6

▶ **Naming acids** A binary acid contains hydrogen and one other element. To name such an acid, use the prefix *hydro-* to name the hydrogen part of the molecule. Then to a form of the root of the second element name, add the suffix *-ic*. Finally, add the word *acid*.

An oxyanion is a polyatomic ion that contains oxygen. An **oxyacid** is an acid that contains hydrogen and an oxyanion. To name such an acid, first write a form of the root of the name of the oxyanion. If the oxyanion name ends in *-ate*, name the acid by adding the suffix *-ic* to the root. If the oxyanion name ends in *-ite*, name the acid by adding the suffix *-ous* to the root. Then add the word *acid*.

Example Problem 9-4
Naming Acids

Name the following acids.

  a. H_2Se

  b. $HBrO_3$

  a. H_2Se is a binary acid, so the prefix *hydro-* is used, followed by the root of the name of the second element (selenium, root *selen-*), followed by the suffix *-ic* and the word *acid*. Thus, the name is hydroselenic acid.

  b. The oxyanion BrO_3^- is the bromate ion. Its root is *brom-*. Because the oxyanion name ends in *-ate*, add the suffix *-ic*. Then add the word *acid*. Thus, the name is bromic acid.

Practice Problems

7. Name the following acids.

    a. H_2CO_3 **b.** HI   **c.** $HClO_4$ **d.** H_2SO_3

▶ **Writing formulas from names** Given the name of a molecular compound, you can write its formula by analyzing the name in terms of the rules given above. Use the prefixes that indicate number (*mono-*, *di-*, and so on) to write the proper subscripts.

Practice Problems

8. Write the formulas for the following molecular compounds.

    a. disulfur dichloride   **c.** hydrosulfuric acid

    b. dinitrogen tetroxide   **d.** sulfuric acid

9.3 Molecular Structures

A **structural formula** is a molecular model that uses letter symbols and bonds to show relative positions of atoms. It can generally be predicted by means of a Lewis structure. The following steps can be used in drawing Lewis structures.

1. Predict the location of certain atoms. Hydrogen is always a terminal atom. The atom with the least attraction for electrons is the central atom.

2. Find the total number of electrons available for bonding (valence electrons). If the structure is to represent a positive or negative polyatomic ion, the ion charge must be subtracted or added, respectively.

3. Divide the total number of available electrons by 2 to obtain the number of bonding pairs.

4. Place one bonding pair between the central atom and each terminal atom.

5. To find the total number of lone pairs and pairs available for multiple bonding, subtract the number of bonding pairs used in step 4 from the number of bonding pairs determined in step 3. Place lone pairs around the terminal atoms to satisfy the octet rule. Assign remaining pairs to the central atom.

6. If the central atom is not surrounded by four electron pairs, convert one or two lone pairs on the terminal atoms to a double or triple bond to the central atom.

Example Problem 9-5
Drawing Lewis Structures

Draw Lewis structures for each of the following.

a. phosphorus trichloride (PCl_3)

b. sulfate ion (SO_4^{2-})

Apply the six steps in each case.

1. The central atoms, which have the least attraction for shared electrons, must be:

   a. P (P is in group 5A, whereas S is in group 7A)

   b. S (S is below O in group 6A)

2. The total available electrons for bonding are:

   a. (5 from P) + (3 × 7 from Cl) = 26

   b. (6 from S) + (4 × 6 from O) + 2 (from charge) = 32

3. Divide the number of total available electrons by 2 to find the number of bonding pairs:

   a. $26/2 = 13$

   b. $32/2 = 16$

4. Place one bonding pair between the central atom and each terminal atom:

   a.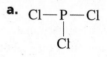

   b.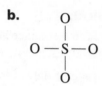

5. Subtract the number of bonding pairs used in step 4 from the number of bonding pairs determined in step 3, and place lone pairs around the terminal atoms to satisfy the octet rule. Assign remaining pairs to the central atom:

   a. $13 - 3 = 10$

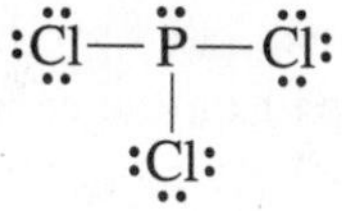

   b. $16 - 4 = 12$

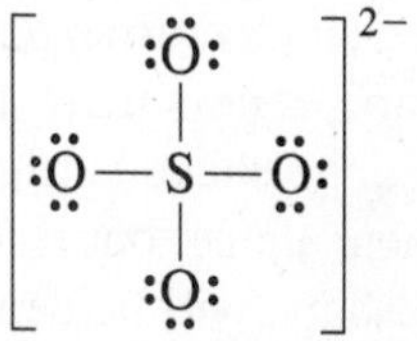

$$:\ddot{Cl}\!-\!\ddot{P}\!-\!\ddot{Cl}: \qquad\qquad \left[\begin{array}{c} :\ddot{O}: \\ | \\ :\ddot{O}\!-\!S\!-\!\ddot{O}: \\ | \\ :\ddot{O}: \end{array}\right]^{2-}$$

6. In each case, the central atom is surrounded by four electron pairs, and all the atoms have an octet. The structures need no further modification.

Practice Problems

9. Draw Lewis structures for each of the following.

   a. CF_4 **b.** CO **c.** SiS_2 **d.** NH_4^+

▶ **Resonance and exceptions to the octet rule** Resonance occurs when more than one valid Lewis structure can be written for a molecule or an ion. For example, three resonance structures exist for the NO_3^- ion because a double bond can be placed between the central N atom and any of the three O atoms.

Sometimes there are exceptions to the octet rule. There may be an odd number of total valence electrons, as in the case of ClO_2. Alternatively, a central atom may have more or fewer than eight electrons. In the latter case, an atom with a lone pair may attach to it by means of a **coordinate covalent bond,** a bond in which both shared electrons are donated by only one of the atoms.

Practice Problems

10. Draw the three resonance structures for the carbonate ion $(CO_3{}^{2-})$. (Hint: Each structure contains one double bond.)

11. State why each of the following is an exception to the octet rule.

   a. NO_2 **b.** BCl_3 **c.** PF_5

9.4 Molecular Shape

The **VSEPR model** is used to determine molecular shape. It assumes arrangements that minimize repulsion of electron pairs around the central atom. The presence of four bonding pairs produces a tetrahedral arrangement. Three such pairs produce a trigonal planar shape, unless there also is a lone pair on the central atom, in which case the shape is trigonal pyramidal. Two bonding pairs produce a linear shape, but if there also are two lone pairs on the central atom, the shape will be bent.

During bonding, atomic orbitals can undergo **hybridization,** or mixing to form new, identical hybrid orbitals. An s orbital and three p orbitals can hybridize to form four identical sp^3 hybrid orbitals. That produces a tetrahedral, trigonal pyramidal, or bent molecular shape, depending on whether none, one, or two of these orbitals contain lone pairs (as in the case of CH_4, NH_3, and H_2O, respectively). The hybridization of an s and two p orbitals produces three sp^2 hybrid orbitals, which produces a trigonal planar shape (as in the case of $AlCl_3$). The hybridization of an s and one p orbital produces two sp hybrid orbitals, which produce a linear shape (as in the case of $BeCl_2$).

Practice Problems

12. Use the VSEPR model and the concept of hybridization to describe the hybrid orbitals and the shape of each of the following molecules.

a. H_2S **d.** NF_3

b. BeF_2 **e.** BI_3

c. CBr_4

9.5 Electronegativity and Polarity

Electronegativity is a measure of the tendency of an atom to attract electrons in a chemical bond. The most highly electronegative elements occur in the upper right in the periodic table (disregarding the noble gases, which rarely form compounds). The least electronegative elements occur in the lower left of the table.

The electronegativity difference between two bonded atoms can be used to predict the character of the bond. A bond is nonpolar covalent when there is virtually no electronegativity difference. A bond is **polar covalent** when there is some electronegativity difference (unequal electron sharing). A bond is ionic if the electronegativity difference is very large. An electronegativity difference of more than 1.7 generally means that a bond is more ionic than covalent.

A molecule also may be polar or nonpolar, depending on whether it contains polar bonds and depending on its shape. For example, a water molecule is polar because there is a significant electronegativity difference between the H and O atoms and because the molecule is bent, not symmetric. There is a negative end near the more electronegative O atom, and a positive end near the H atoms. Carbon tetrachloride (CCl_4) has polar bonds, but the tetrahedral molecule as a whole is nonpolar because it is symmetric.

The properties of covalent substances vary, depending on a number of factors, including the intermolecular forces. Polar molecules tend to be soluble in polar solvents, whereas nonpolar molecules tend to be soluble in nonpolar solvents. For nonpolar molecules, the intermolecular forces are weak dispersion forces, and melting and boiling points tend to be low. For polar molecules, there are stronger dipole-dipole forces whose magnitude depends on the degree of polarity. If hydrogen is bonded to a highly electronegative atom, such as oxygen, the intermolecular force can be quite strong and is called hydrogen bonding. Such covalent substances have relatively high boiling points for their molecular size. In some covalent substances, the atoms are connected in large networks of bonds. Such substances, called covalent network solids, are very hard and have very high melting points.

Practice Problems

13. Use the electronegativity values in Figure 9-15 in your textbook to determine whether the following molecules have polar or nonpolar bonds. Apply the VSEPR model to determine the shape of each molecule. Then state whether the molecule as a whole is polar or nonpolar.

   a. SiF_4 **c.** H_2S

   b. PCl_3 **d.** CH_3Cl

Chapter 9 Review

14. Why do atoms tend to bond?

15. Define *covalent bond*.

16. What is the difference between a sigma bond and a pi bond? How many of each type of bond are found in a triple bond?

17. What is bond dissociation energy?

18. What kind of reaction occurs when a greater amount of energy is required to break the existing bonds in the reactants than is released when the new bonds form in the products?

19. State the rules for naming binary molecular compounds.

20. State the steps used to draw a Lewis structure.

21. Explain when the condition called resonance occurs.

22. What is the VSEPR model used to determine?

23. Define *hybridization*.

24. When is a bond polar? When is a molecule polar?

Chemical Reactions

10.1 Reactions and Equations

The process by which atoms of one or more substances are rearranged to form different substances is called a **chemical reaction.** There are a number of possible indications that a chemical reaction has occurred. Among these are absorption of heat and release of energy in the form of light or heat, color change, production of an odor, and appearance of gas bubbles or a solid.

Statements called equations are used to represent chemical reactions. Equations indicate the **reactants,** or starting substances, and the **products,** or substances formed during the reaction. The reactants are written to the left of an arrow that stands for "yields," and the products are written to the right. Plus signs are used to separate the different reactants or products. The symbols (s), (l), (g), and (aq) indicate solid, liquid, gas, or water solution (aqueous), respectively.

Equations can be in the form of word equations, in which the reactants or products are indicated by their names. Word equations can be replaced by skeleton equations, which use chemical formulas rather than words to identify reactants and products. Skeleton equations are more informative than word equations because skeleton equations identify the atoms that make up each reactant and product.

Example Problem 10-1
Writing Word and Skeleton Equations

Write a word equation and a skeleton equation for the reaction in which hydrogen gas reacts with oxygen gas to form liquid water.

The word equation uses the names and physical states of the reactants and products, along with the arrow and plus symbols.

$$\text{hydrogen(g)} + \text{oxygen(g)} \rightarrow \text{water(l)}$$

The skeleton equation replaces the names of the substances with their chemical formulas.

$$H_2(g) + O_2(g) \rightarrow H_2O(l)$$

Practice Problems

1. Write a word equation and a skeleton equation for each of the following descriptions of chemical reactions.

   a. Solid lithium reacts with chlorine gas to produce solid lithium chloride.

   b. Nitrogen gas reacts with oxygen gas to produce nitrogen dioxide gas.

   c. Solid iron reacts with aqueous copper(II) nitrate to produce solid copper and aqueous iron(II) nitrate.

▶ **Balanced chemical equations** Skeleton equations do not reflect the fact that matter is conserved during a reaction. In actual chemical reactions, atoms are neither created nor destroyed but are conserved. An equation that reflects the fact that the same number of each kind of atom must appear on both sides of the arrow is called a balanced **chemical equation.** It uses chemical formulas to show the identities and relative amounts of the substances involved in a reaction.

To balance a chemical equation, you write **coefficients,** which are numbers placed in front of the chemical formulas. The lowest whole-number ratio is used. If no coefficient appears next to a formula, the number 1 is understood. The steps involved in balancing a chemical equation are as follows.

Step 1 Write the skeleton equation.

Step 2 Count the atoms of the elements in the reactants. Any polyatomic ions that remain unchanged can be counted as if they are elements.

Step 3 Count the atoms of the elements in the products.

Step 4 Change the coefficients to make the number of atoms of each element equal on both sides of the equation. Do not change subscripts.

Step 5 Write the coefficients in their lowest possible ratio.

Step 6 Check your work.

Example Problem 10-2
Writing a Balanced Chemical Equation

Write the balanced chemical equation for the reaction in which solid
sulfur (S) combines with oxygen gas to produce sulfur trioxide gas.

Step 1 Write the skeleton equation.

$$S(s) + O_2(g) \rightarrow SO_3(g)$$

Step 2 Count atoms of each element in the reactants.

$$1\ S, 2\ O$$

Step 3 Count atoms of each element in the products.

$$1\ S, 3\ O$$

Step 4 Adjust coefficients.
In the formulas O_2 and SO_3, oxygen is in a 2 to 3 ratio. To balance
O, 3 O_2 are needed for every 2 SO_3. Insert the coefficients 3 and 2,
respectively.

$$S(s) + 3O_2(g) \rightarrow 2SO_3(g)$$

Now the S atoms must be accounted for. The placement of the
2 before the SO_3 means that there are now 2 S atoms on the right.
There must also be 2 on the left. Place the coefficient 2 in front of
the reactant S.

$$2S(s) + 3O_2(g) \rightarrow 2SO_3(g)$$

Step 5 Write the coefficients in their lowest possible ratio.
The ratio 2:3:2 cannot be written any lower and still produce whole
numbers. The ratio is already the lowest possible.

Step 6 Check your work.
Count the number of atoms of each element on both sides of the
equation and make sure they are equal.
Reactants: 2 S, 6 O   Products: 2 S, 6 O
The numbers are equal, so the equation is correctly balanced.

Practice Problems

2. Write a balanced chemical equation for the following reactions, making sure coefficients are in their lowest possible ratio.

   a. Solid potassium reacts with liquid water to produce hydrogen gas and a water solution of potassium hydroxide.

   b. Calcium chloride and sodium carbonate in water solution produce solid calcium carbonate and a water solution of sodium chloride.

   c. Liquid bromine and solid lithium iodide react to form solid lithium bromide and solid iodine.

10.2 Classifying Chemical Reactions

Reactions can be classified into various types. In a **synthesis reaction,** two or more substances react to produce a single product. An example is the reaction between the elements potassium and bromine to form potassium bromide.

$$2K(s) + Br_2(l) \rightarrow 2KBr(s)$$

Synthesis reactions can also occur between compounds or between a compound and an element. In a **combustion reaction,** oxygen combines with a substance and produces light and heat. The burning of methane in oxygen is an example of a combustion reaction.

$$CH_4(g) + 2O_2(g) \rightarrow CO_2(g) + 2H_2O(g)$$

Certain reactions may be examples of both synthesis and combustion. An example is the burning of carbon.

$$C(s) + O_2(g) \rightarrow CO_2(g)$$

A **decomposition reaction** is one in which a single compound breaks down into two or more elements or new compounds. An example is the breakdown of aluminum oxide into aluminum and oxygen gas.

$$2Al_2O_3(s) \rightarrow 4Al(s) + 3O_2(g)$$

Practice Problems

3. Identify each of the following skeleton equations as representing a synthesis reaction, a combustion reaction, both synthesis and combustion, or decomposition reaction. Balance each equation if necessary.

a. $N_2(g) + H_2(g) \rightarrow NH_3(g)$

b. $CaCO_3(s) \rightarrow CaO(s) + O_2(g)$

c. $Se(s) + O_2(g) \rightarrow SeO_3(g)$

d. $C_2H_4(g) + O_2(g) \rightarrow CO_2(g) + H_2O(g)$

▶ **Replacement reactions** In a **single-replacement reaction,** atoms of an element replace the atoms of another element in a compound. An example of such a reaction is the replacement by zinc metal of silver in silver nitrate water solution.

$$Zn(s) + 2AgNO_3(aq) \rightarrow Zn(NO_3)_2(aq) + 2Ag(s)$$

Not all metals will replace one another. Only a metal that is more reactive chemically will replace a chemically less reactive metal. The order of such reactivity is called an activity series, as shown below.

Most active
Lithium
Potassium
Sodium
Magnesium
Aluminum
Zinc
Iron
Lead
Copper
Least active Silver

Example Activity Series

Any metal in an activity series will replace the metals below it in their compounds. In a similar fashion, any elemental halogen will replace another halogen from its compounds if the second halogen lies below it in the periodic table. Thus, fluorine is more active than chlorine, which is more active than bromine, which is more active than iodine.

Example Problem 10-3
Using an Activity Series to Predict Single-Replacement Reactions

Use the above activity series to predict the single-replacement reaction that will occur between aluminum metal and lead(II) nitrate $(Pb(NO_3)_2)$ in a water solution and write its balanced chemical equation.

Aluminum is above lead in the activity series and therefore is more active. It will replace lead in its compounds in water solution. A single-replacement reaction will take place to form the nitrate of aluminum, a group 3A metal. The skeleton equation is shown below.

$$Al(s) + Pb(NO_3)_2(aq) \rightarrow Al(NO_3)_3(aq) + Pb(s)$$

Adding coefficients, the balanced chemical equation is shown below.

$$2Al(s) + 3Pb(NO_3)_2(aq) \rightarrow 2Al(NO_3)_3(aq) + 3Pb(s)$$

Practice Problems

4. Predict whether a single-replacement reaction will occur between the following pairs of possible reactants. If so, write a balanced chemical equation for the reaction.

   a. chlorine gas and aqueous potassium iodide

   b. magnesium metal and aqueous copper(II) sulfate

   c. copper metal and iron(III) chloride in water solution

   d. lead metal and aqueous silver nitrate

In another type of replacement reaction, called a **double-replacement reaction,** there is an exchange of positive ions between two compounds, often in water solution. Often, a solid of low solubility is produced during the reaction and settles out of the solution. Such a solid is called a **precipitate.** An example of a double-replacement reaction that produces the precipitate lead(II) chloride is the following.

$$Pb(NO_3)_2(aq) + 2NaCl(aq) \rightarrow 2NaNO_3(aq) + PbCl_2(s)$$

Practice Problems

5. Write a balanced chemical equation for each of the following double-replacement reactions.

a. A water solution of zinc bromide ($ZnBr_2$) and a water solution of potassium hydroxide form a water solution of potassium bromide and a precipitate of zinc hydroxide.

b. A water solution of copper(II) sulfate and a water solution of barium chloride produce a water solution of copper(II) chloride and solid barium sulfate.

c. A precipitate of iron(III) carbonate and a water solution of sodium nitrate are formed when a water solution of iron(III) nitrate and a water solution of sodium carbonate are mixed.

10.3 Reactions in Aqueous Solutions

When a substance dissolves in water, a solution forms. A solution is a homogeneous mixture because it has a constant composition throughout. A solution contains one or more substances called **solutes** dissolved in the solvent. The **solvent** is the most plentiful substance in a solution. An **aqueous solution** is a solution in which the solvent is water. When dissolved to form aqueous solutions, the ions of ionic compounds separate. Some molecular compounds also produce dissolved ions in water. If hydrogen ions are produced, the substance is called an acid. For example, the gaseous molecular compound hydrogen chloride (HCl) forms H^+ and Cl^- ions in aqueous solution, which is called hydrochloric acid.

When aqueous solutions that contain ions are mixed, the ions may react in a double-replacement reaction. The product is typically a solid precipitate, water, or a gas. An example of a double-replacement reaction that produces a precipitate occurs when aqueous solutions of sodium chloride and silver nitrate are mixed to form a precipitate of solid silver chloride.

$$NaCl(aq) + AgNO_3(aq) \rightarrow NaNO_3(aq) + AgCl(s)$$

To show all of the particles in solution as they really exist, a **complete ionic equation** can be written.

$$Na^+(aq) + Cl^-(aq) + Ag^+(aq) + NO_3^-(aq) \rightarrow$$
$$Na^+(aq) + NO_3^-(aq) + AgCl(s)$$

The sodium and nitrate ions are on both sides of the equation. Such ions that do not participate in a reaction are called **spectator ions.** An ionic equation that does not show spectator ions but only the particles that participate in a reaction is called a **net ionic equation.** In the case of the reaction above, the net ionic equation from which the sodium and nitrate ions have been removed is as follows.

$$Cl^-(aq) + Ag^+(aq) \rightarrow AgCl(s)$$

Example Problem 10-4
Writing Ionic Equations

Write the balanced chemical equation for the reaction between aqueous solutions of strontium nitrate and potassium sulfate, which forms the precipitate strontium sulfate. Then write the complete ionic and net ionic equations.

Write the correct skeleton equation.

$$Sr(NO_3)_2(aq) + K_2SO_4(aq) \rightarrow KNO_3(aq) + SrSO_4(s)$$

Use coefficients to produce the balanced chemical equation.

$$Sr(NO_3)_2(aq) + K_2SO_4(aq) \rightarrow 2KNO_3(aq) + SrSO_4(s)$$

Write the complete ionic equation.

$$Sr^{2+}(aq) + 2NO_3^-(aq) + 2K^+(aq) + SO_4^{2-}(aq) \rightarrow$$
$$2K^+(aq) + 2NO_3^-(aq) + SrSO_4(s)$$

Cross out the spectator ions, which are those that are on both sides of the equation.

$$Sr^{2+}(aq) + \cancel{2NO_3^-(aq)} + \cancel{2K^+(aq)} + SO_4^{2-}(aq) \rightarrow$$
$$\cancel{2K^+(aq)} + \cancel{2NO_3^-(aq)} + SrSO_4(s)$$

That leaves the net ionic equation.

$$Sr^{2+}(aq) + SO_4^{2-}(aq) \rightarrow SrSO_4(s)$$

Practice Problems

6. Write balanced chemical, complete ionic, and net ionic equations for each of the following reactions.

   a. Aqueous solutions of lead(II) nitrate and ammonium chloride are mixed, forming a precipitate of lead(II) chloride.

   b. Aqueous solutions of aluminum chloride and sodium carbonate are combined, producing solid aluminum carbonate.

▶ **Reactions that form water or a gas** Some double-replacement reactions in aqueous solution produce water or a gas (or both) rather than a precipitate. In such cases, the water or gas is shown as a product in the net ionic equation, as are the ions that produced it. The remaining ions are eliminated as spectator ions. The following example problem illustrates this concept.

Example Problem 10-5
Writing Equations for a Reaction That Produces Water

When hydrochloric acid and potassium hydroxide solutions are mixed, water results, together with an aqueous solution of potassium chloride. Write the balanced chemical equation, a complete ionic equation, and a net ionic equation for this reaction.

The balanced chemical equation is the same as the skeleton equation.

$$HCl(aq) + KOH(aq) \rightarrow H_2O(l) + KCl(aq)$$

Write the complete ionic equation, which includes all of the ions.

$$H^+(aq) + Cl^-(aq) + K^+(aq) + OH^-(aq) \rightarrow$$
$$H_2O(l) + K^+(aq) + Cl^-(aq)$$

Remove the spectator ions to produce the net ionic equation.

$$H^+(aq) + OH^-(aq) \rightarrow H_2O(l)$$

Practice Problems
7. Write balanced chemical, complete ionic, and net ionic equations for the reactions between the following substances, which produce water.
   a. nitric acid (HNO_3) and aqueous barium hydroxide
   b. sulfuric acid (H_2SO_4) and aqueous sodium hydroxide
   c. phosphoric acid (H_3PO_4) and aqueous lithium hydroxide

Example Problem 10-6
Writing Equations for a Reaction That Produces a Gas

Write balanced chemical, complete ionic, and net ionic equations for the reaction between aqueous solutions of sodium sulfide and hydrochloric acid. Hydrogen sulfide gas is produced, along with an aqueous solution of sodium chloride.

Write the skeleton equation with the correct formulas.

$$Na_2S(aq) + HCl(aq) \rightarrow NaCl(aq) + H_2S(g)$$

Add coefficients to write a balanced chemical equation.

$$Na_2S(aq) + 2HCl(aq) \rightarrow 2NaCl(aq) + H_2S(g)$$

Write the complete ionic equation, which includes all the ions.

$$2Na^+(aq) + S^{2-}(aq) + 2H^+(aq) + 2Cl^-(aq) \rightarrow$$
$$2Na^+(aq) + 2Cl^-(aq) + H_2S(g)$$

Remove the spectator ions to produce the net ionic equation.

$$S^{2-}(aq) + 2H^+(aq) \rightarrow H_2S(g)$$

Practice Problems

8. Write balanced chemical, complete ionic, and net ionic equations for the reactions between the following substances, which produce a gas.

  a. hydrochloric acid and aqueous sodium cyanide, with production of hydrogen cyanide gas (HCN)

  b. sulfuric acid (H_2SO_4) and aqueous rubidium sulfide, with production of hydrogen sulfide gas

Chapter 10 Review

9. Define *chemical reaction*. Where are the reactants and products shown in an equation for a chemical reaction, and how are the physical states indicated?

10. Compare and contrast word equations, skeleton equations, and balanced chemical equations.

11. Briefly list the steps used in balancing a chemical equation.

12. Contrast synthesis, combustion, and decomposition reactions.

13. Contrast single-replacement and double-replacement reactions.

14. Define *aqueous solution*. What name is given to a solid that forms when two aqueous solutions are mixed?

15. How do complete ionic and net ionic equations differ?

16. Name three typical types of products that may be formed in double-replacement reactions.

The Mole

11.1 Measuring Matter

Substances react according to definite ratios of numbers of particles (atoms, ions, formula units, or molecules). The following balanced chemical equation shows that two atoms of aluminum react with three molecules of iodine to form two formula units of aluminum iodide.

$$2Al(s) + 3I_2(s) \rightarrow 2AlI_3(s)$$

If you wanted to carry out this reaction, how could you measure the correct numbers of particles of aluminum and iodine? Fortunately, you can count particles by measuring mass. Suppose you have a sample of an element, and the mass of the sample in grams is numerically equal to the atomic mass of the element. Scientists have discovered that this mass of an element contains 6.02×10^{23} atoms of that element. This number is called **Avogadro's number.** Avogadro's number of particles is called a **mole** of particles. The mole is the SI base unit used to measure the amount of a substance and is defined as the number of particles in exactly 12 g of pure carbon-12. For the purpose of dealing with moles, the simplest unit of any substance is called a representative particle. The particle may be an atom, a molecule, a formula unit, or an ion.

Example Problem 11-1
Converting Moles to Number of Particles

How many molecules are in 2.25 moles of bromine (Br_2)?

Bromine is an element that consists of diatomic molecules. Therefore, one mole of bromine contains 6.02×10^{23} Br_2 molecules. To find the number of Br_2 molecules present in 2.25 mol, multiply the number of moles by Avogadro's number.

$$\text{number of } Br_2 \text{ molecules} =$$

$$2.25 \ \cancel{mol} \times \frac{6.02 \times 10^{23} \text{ molecules } Br_2}{1 \ \cancel{mol}}$$

$$\text{number of } Br_2 \text{ molecules} = 1.35 \times 10^{24} \text{ molecules } Br_2$$

Example Problem 11-2

Converting Number of Particles to Moles

Calculate the number of moles in a sample of sodium bromide (NaBr) that contains 2.88×10^{23} formula units.

Because 1 mol NaBr = 6.02×10^{23} formula units NaBr, you can see that 2.88×10^{23} formula units is less than one mole of NaBr.

$$\text{moles of NaBr} =$$

$$2.88 \times 10^{23} \text{ formula units} \times \frac{1 \text{ mol NaBr}}{6.02 \times 10^{23} \text{ formula units}}$$

$$\text{moles of NaBr} = 0.478 \text{ mol NaBr}$$

Practice Problems

1. Calculate the number of molecules in 15.7 mol carbon dioxide.
2. Calculate the number of molecules in 0.0544 mol H_2O.
3. Calculate the number of moles in 9.22×10^{23} atom iron.
4. Calculate the number of moles of sucrose in a sample that contains 2.05×10^{22} sucrose molecules.
5. A student uses 0.0850 mol copper sulfate to carry out a reaction. If the reaction uses up 0.0832 mol copper sulfate, how many formula units are left unreacted?

11.2 Mass and the Mole

One mole of a monatomic element consists of 6.02×10^{23} atoms of that element. The mass of a mole of any substance is called the **molar mass** of the substance. For example, the molar mass of a monatomic element is numerically equal to the atomic mass of the element, but expressed in grams.

$$\text{molar mass of a substance} = \frac{\text{grams of the substance}}{1 \text{ mol of the substance}}$$

This relationship can be used to convert between mass and moles.

Example Problem 11-3
Converting Mass to Moles

A roll of copper wire has a mass of 848 g. How many moles of copper are in the roll? Use the atomic mass of copper given on the periodic table to apply a conversion factor to the mass given in the problem.

$$\text{moles of Cu} = \text{grams Cu} \times \frac{1 \text{ mol Cu}}{\text{grams Cu}}$$

$$\text{moles of Cu} = 848 \text{ g Cu} \times \frac{1 \text{ mol Cu}}{63.546 \text{ g Cu}} = 13.3 \text{ mol Cu}$$

Note that the conversion factor is arranged so that g Cu cancel, leaving only mol Cu, the desired quantity.

Example Problem 11-4
Converting Moles to Mass

Calculate the mass of 0.625 moles of calcium.

Use the molar mass of calcium to apply a conversion factor to the number of moles given in the problem. According to the periodic table, the atomic mass of calcium is 40.078 amu, so the molar mass of calcium is 40.078 g.

$$\text{mass of Ca} = \text{mol Ca} \times \frac{\text{grams Ca}}{1 \text{ mol ca}}$$

$$\text{mass of Ca} = 0.625 \text{ mol Ca} \times \frac{40.078 \text{ g Ca}}{1 \text{ mol Ca}} = 25.0 \text{ g Ca}$$

The conversion factor is arranged so that mol Ca cancel, leaving g Ca, the desired quantity.

Practice Problems

6. Calculate the mass of 6.89 mol antimony.

7. A chemist needs 0.0700 mol selenium for a reaction. What mass of selenium should the chemist use?

8. A sample of sulfur has a mass of 223 g. How many moles of sulfur are in the sample?

9. A tank of compressed helium contains 50.0 g helium. How many moles of helium are in the tank?

10. A nickel coin is 25.0% nickel and 75.0% copper. If a nickel has a mass of 5.00 g, how many moles of Ni are in a nickel coin?

Now that you can relate mass to moles, you can also relate mass to number of particles by first converting mass to moles and then using the following relationship.

1 mol of a substance = 6.02×10^{23} particles of the substance

Example Problem 11-5
Converting Mass to Number of Particles

Calculate the number of atoms in 4.77 g lead.

To find the number of atoms in the sample, you must first determine how many moles are in 4.77 g lead. According to data from the periodic table, the molar mass of lead is 207.2 g/mol. Apply a conversion factor to convert mass to moles.

$$\text{moles of Pb} = \text{grams Pb} \times \frac{1 \text{ mol Pb}}{\text{grams Pb}}$$

$$\text{moles of Pb} = 4.77 \text{ g Pb} \times \frac{1 \text{ mol Pb}}{207.2 \text{ g Pb}} = 0.0230 \text{ mol Pb}$$

Now use a second conversion factor to convert moles to number of particles.

$$\text{atoms of Pb} = \text{mol Pb} \times \frac{6.02 \times 10^{23} \text{ atom Pb}}{1 \text{ mol Pb}}$$

$$\text{atoms of Pb} = 0.0230 \text{ mol Pb} \times \frac{6.02 \times 10^{23} \text{ atom Pb}}{1 \text{ mol Pb}} =$$

$$1.38 \times 10^{22} \text{ atom Pb}$$

You can also convert from number of particles to mass by reversing the procedure above and dividing the number of particles by Avogadro's number to determine the number of moles present.

Practice Problems

11. Calculate the number of atoms in 2.00 g of platinum.

12. How many sulfur atoms are in a metric ton (1.00×10^6 g) of sulfur?

13. How many grams of mercury are in 1.19×10^{23} atoms of mercury?

14. What is the mass in grams of 3.01×10^{19} atoms of iodine?

15. The EPA limit for lead in the water supply is 15 parts per billion by mass. Calculate the number of lead ions present in 1.00 kg of water that is at the EPA limit for lead.

11.3 Moles of Compounds

Recall that a mole is Avogadro's number (6.02×10^{23}) of particles of a substance. If the substance is a molecular compound, such as ammonia (NH_3), a mole is 6.02×10^{23} molecules of ammonia. If the substance is an ionic compound, such as baking soda (sodium hydrogen carbonate, $NaHCO_3$), a mole is 6.02×10^{23} formula units of sodium hydrogen carbonate. In either case, a mole of a compound contains as many moles of each element as are indicated by the subscripts in the formula for the compound. For example, a mole of ammonia (NH_3) consists of one mole of nitrogen atoms and three moles of hydrogen atoms.

▶ **Molar mass of a compound** The molar mass of a compound is the mass of a mole of the representative particles of the compound. Because each representative particle is composed of two or more atoms, the molar mass of the compound is found by adding the molar masses of all of the atoms in the representative particle. In the case of NH_3, the molar mass equals the mass of one mole of nitrogen atoms plus the mass of three moles of hydrogen atoms.

molar mass of NH_3 = molar mass of N + 3(molar mass of H)

molar mass of NH_3 = 14.007 g + 3(1.008 g) = 17.031 g/mol

You can use the molar mass of a compound to convert between mass and moles, just as you used the molar mass of elements to make these conversions.

Example Problem 11-6
Converting Mass of a Compound to Moles

At 4.0°C, water has a density of 1.000 g/mL. How many moles of water are in 1.000 kg of water (1.000 L at 4.0°C)?

Before you can calculate moles, you must determine the molar mass of water (H_2O). A mole of water consists of two moles of hydrogen atoms and one mole of oxygen atoms.

$$\text{molar mass } H_2O = 2(\text{molar mass H}) + \text{molar mass O}$$

$$\text{molar mass } H_2O = 2(1.008 \text{ g}) + 15.999 \text{ g} = 18.015 \text{ g/mol}$$

Now you can use the molar mass of water as a conversion factor to determine moles of water. Notice that 1.000 kg is converted to 1.000×10^3 g for the calculation.

$$\text{moles of } H_2O = \text{grams } H_2O \times \frac{1 \text{ mol } H_2O}{18.015 \text{ g } H_2O}$$

$$\text{moles of } H_2O = 1.000 \times 10^3 \text{ g } H_2O \times \frac{1 \text{ mol } H_2O}{18.015 \text{ g } H_2O} =$$

$$0.05551 \times 10^3 \text{ mol } H_2O$$

$$\text{moles of } H_2O = 55.51 \text{ mol } H_2O$$

Notice that the result must have four significant figures because the mass and volume data in the problem were given to four significant figures.

Practice Problems

16. Calculate the number of moles in 17.2 g of benzene (C_6H_6).
17. Calculate the number of moles in 350.0 g of potassium chlorate ($KClO_3$).
18. Determine the mass of 0.187 mol of tin(II) sulfate ($SnSO_4$).
19. A chemist needs 1.35 mol of ammonium dichromate for a reaction. The formula for this substance is $(NH_4)_2Cr_2O_7$. What mass of ammonium dichromate should the chemist measure out?
20. A student needs 0.200 mol each of zinc metal and copper(II) nitrate ($Cu(NO_3)_2$) for an experiment. What mass of each should the student obtain?

11.4 Empirical and Molecular Formulas

Recall that every chemical compound has a definite composition—a composition that is always the same wherever that compound is found. The composition of a compound is usually stated as the percent by mass of each element in the compound.

▶ **Percent composition** The percent of an element in a compound can be found in the following way.

$$\% \text{ by mass of an element} =$$

$$\frac{\text{mass of element in 1 mol compound}}{\text{molar mass of compound}} \times 100$$

The example problem below shows you how to determine the **percent composition** of a compound, which is the percent by mass of each element in the compound.

Example Problem 11-7
Calculating Percent Composition

Determine the percent composition of calcium chloride ($CaCl_2$).

First, analyze the information available from the formula. A mole of calcium chloride consists of one mole of calcium ions and two moles of chloride ions.

Next, gather molar mass information from the atomic masses on the periodic table. To the mass of one mole of $CaCl_2$, a mole of calcium ions contributes 40.078 g, and two moles of chloride ions contribute 2×35.453 g = 70.906 g for a total molar mass of 110.984 g/mol for $CaCl_2$.

Finally, use the data to set up a calculation to determine the percent by mass of each element in the compound. The percent by mass of calcium and chlorine in $CaCl_2$ can be calculated as follows.

$$\% \text{ Ca in Ca Cl}_2 = \frac{40.078 \text{ g Ca}}{110.984 \text{ g CaCl}_2} \times 100 = 36.112\% \text{ Ca}$$

$$\% \text{ Cl in CaCl}_2 = \frac{70.906 \text{ g Cl}}{110.984 \text{ g CaCl}_2} \times 100 = 63.888\% \text{ Cl}$$

As a check, be sure that the percentages add up to 100%. In this case, the percentages add up to 100.000%.

Practice Problems

21. Calculate the percent composition of aluminum oxide (Al_2O_3).

22. Determine the percent composition of magnesium nitrate, which has the formula $Mg(NO_3)_2$.

23. Calculate the percent oxygen in potassium chlorate ($KClO_3$).

24. Calculate the percent nitrogen in ammonium hexacyanoiron(II), which has the formula $(NH_4)_4Fe(CN)_6$.

25. Acetylene gas has the molecular formula C_2H_2. How does the percent composition of acetylene compare with that of benzene (C_6H_6)?

▶ **Empirical formulas** You can use percent composition data to help identify an unknown compound by determining its empirical formula. The **empirical formula** is the simplest whole-number ratio of atoms of elements in the compound. In many cases, the empirical formula is the actual formula for the compound. For example, the simplest ratio of atoms of sodium to atoms of chlorine in sodium chloride is 1 atom Na : 1 atom Cl. So, the empirical formula of sodium chloride is Na_1Cl_1, or NaCl, which is the true formula for the compound. The following example problem will show you how to determine empirical formulas.

Example Problem 11-8
Empirical Formula from Percent Composition

The percent composition of an unknown compound is found to be 38.43% Mn, 16.80% C, and 44.77% O. Determine the compound's empirical formula.

Because *percent* means "parts per hundred parts," assume that you have 100 g of the compound. Then calculate the number of moles of each element in the 100 g of compound. The number of moles of manganese may be calculated as follows.

$$\text{moles of Mn (in 100 g)} = \text{grams Mn (in 100 g)} \times \frac{1 \text{ mol Mn}}{\text{molar mass Mn}}$$

$$\text{moles of Mn (in 100 g)} = 38.43 \text{ g Mn} \times \frac{1 \text{ mol Mn}}{54.938 \text{ g Mn}} =$$

$$0.6995 \text{ mol Mn}$$

By following the same pattern, the number of moles of carbon and oxygen per 100-g sample may be calculated.

$$\text{moles of C (in 100 g)} = 16.80 \text{ g C} \times \frac{1 \text{ mol C}}{12.011 \text{ g C}} = 1.399 \text{ mol C}$$

$$\text{moles of O (in 100 g)} = 44.77 \text{ g O} \times \frac{1 \text{ mol O}}{15.999 \text{ g O}} = 2.798 \text{ mol O}$$

The results show the following relationship.

$$\text{mol Mn : mol C : mol O} = 0.6995 : 1.339 : 2.798$$

To obtain the simplest whole-number ratio of moles, divide each number of moles by the smallest number of moles.

$$\text{moles of Mn} = \frac{0.6995 \text{ mol Mn}}{0.6995} = 1 \text{ mol Mn}$$

$$\text{moles of C} = \frac{1.399 \text{ mol C}}{0.6995} = 2 \text{ mol C}$$

$$\text{moles of O} = \frac{2.798 \text{ mol O}}{0.6995} = 4 \text{ mol O}$$

The empirical formula for the compound is MnC_2O_4.

Practice Problems

26. The composition of acetic acid is 40.00% carbon, 6.71% hydrogen, and 53.29% oxygen. Calculate the empirical formula for acetic acid.

27. An unknown compound is analyzed and found to be composed of 14.79% nitrogen, 50.68% oxygen, and 34.53% zinc. Calculate the empirical formula for the compound.

28. Calculate the empirical formula for a compound whose analysis is 74.97% aluminum and 25.03% carbon.

29. The composition of ascorbic acid (vitamin C) is 40.92% carbon, 4.58% hydrogen, and 54.50% oxygen. What is the empirical formula for vitamin C?

▶ **Molecular formulas** For many compounds, the empirical formula is not the true formula. For example, you found the empirical formula for acetic acid in Practice Problem 26 to be CH_2O. Chemists have learned, though, that acetic acid is a molecule with the formula $C_2H_4O_2$, which is the molecular formula for acetic acid. A **molecular formula** tells the exact number of atoms of each element in a molecule or formula unit of a compound. Notice that the molecular formula for acetic acid ($C_2H_4O_2$) has exactly twice as many atoms of each element as the empirical formula (CH_2O). The molecular formula for a compound is always a whole-number multiple of the empirical formula.

In order to determine the molecular formula for an unknown compound, you must know the molar mass of the compound in addition to its empirical formula. Then you can compare the molar mass of the compound with the molar mass represented by the empirical formula as shown in the following example problem.

Example Problem 11-9
Determining a Molecular Formula

Maleic acid is a compound that is widely used in the plastics and textiles industries. The composition of maleic acid is 41.39% carbon, 3.47% hydrogen, and 55.14% oxygen. Its molar mass is 116.1 g/mol. Calculate the molecular formula for maleic acid.

Start by determining the empirical formula for the compound as shown in Example Problem 11-8.

$$\text{moles of C (in 100 g)} = 41.39\ \text{g C} \times \frac{1\ \text{mol C}}{12.011\ \text{g C}} = 3.446\ \text{mol C}$$

$$\text{moles of H (in 100 g)} = 3.47\ \text{g H} \times \frac{1\ \text{mol H}}{1.008\ \text{g H}} = 3.442\ \text{mol H}$$

$$\text{moles of O (in 100 g)} = 55.14\ \text{g O} \times \frac{1\ \text{mol O}}{15.999\ \text{g O}} = 3.446\ \text{mol O}$$

The numbers of moles of C, H, and O are nearly equal, so it is not necessary to divide through by the smallest value. You can see by inspection that the smallest whole-number ratio is 1C : 1H : 1O, and the empirical formula is CHO.

Next, calculate the molar mass represented by the formula CHO. Here, the molar mass is the sum of the masses of one mole of each element.

$$\text{molar mass CHO} = 12.011 \text{ g} + 1.008 \text{ g} + 15.999 \text{ g}$$

$$\text{molar mass CHO} = 29.018 \text{ g/mol}$$

As stated in the problem, the molar mass of maleic acid is known to be 116.1 g/mol. To determine the molecular formula for maleic acid, calculate the whole number multiple, n, to apply to its empirical formula.

$$n = \frac{116.1 \text{ g/mol maleic acid}}{29.018 \text{ g/mol CHO}} = 4.001$$

This calculation shows that the molar mass of maleic acid is four times the molar mass of its empirical formula CHO. Therefore, the molecular formula must have four times as many atoms of each element as the empirical formula. Thus, the molecular formula is (CHO)4 = $C_4H_4O_4$. A check of the molecular formula for maleic acid in a reference book will confirm this result.

Practice Problems

30. Ricinine is one of the poisonous compounds found in the castor plant. The composition of ricinine is 58.54% carbon, 4.91% hydrogen, 17.06% nitrogen, and 19.49% oxygen. Ricinine's molar mass is 164.16 g/mol. Determine its molecular formula.

31. The compound borazine consists of 40.29% boron, 7.51% hydrogen, and 52.20% nitrogen, and its molar mass is 80.50 g/mol. Calculate the molecular formula for borazine.

32. The composition of silver oxalate is 71.02% silver, 7.91% carbon, and 21.07% oxygen. If the molar mass of silver oxalate is 303.8 g/mol, what is its molecular formula?

33. A compound of phosphorus and sulfur contains 27.87% phosphorus and 72.13% sulfur. The molar mass of the compound is 222.3 g/mol. Calculate its molecular formula.

34. Triethylenemelamine is used in the plastics industry and as an anticancer drug. Its analysis is 52.93% carbon, 5.92% hydrogen, and 41.15% nitrogen. The molar mass of triethylenemelamine is 204.2 g/mol. Determine its molecular formula.

11.5 The Formula for a Hydrate

Many compounds, particularly ionic compounds, incorporate a specific number of water molecules into their crystals when they crystallize from aqueous solution. Compounds that include water in their crystal structures are called **hydrates.** The formula for a hydrate is written by adding the formula for water to the formula for the compound. The number of water molecules included per formula unit is indicated by a coefficient. The formula for the blue crystals of copper(II) sulfate is $CuSO_4 \cdot 5H_2O$, indicating that the compound includes five water molecules per formula unit of $CuSO_4$.

Everything you learned earlier in this chapter may be applied to hydrates. The most important thing to remember when dealing with hydrates is that the water molecules are treated as completely separate units—as if the water molecules were atoms of another element in the compound. When giving the percent composition of hydrates, the percent water is given separately, even if the compound contains other oxygen or hydrogen atoms. The following example problem shows you how to determine the number of water molecules included in a hydrate.

Example Problem 11-10
Determining the Formula of a Hydrate

A hydrate of aluminum bromide is composed of 71.16% $AlBr_3$ and 28.84% H_2O. What is the formula for the hydrate?

Again, when dealing with percent composition, assume you are working with a 100-g sample of the compound. A 100-g sample of this hydrate would consist of 71.16 g of $AlBr_3$ and 28.84 g of H_2O. By determining the ratio of moles of $AlBr_3$ to moles of H_2O in this sample, you can write the formula for the hydrate.

First, calculate the moles of $AlBr_3$ and the moles of H_2O. Before starting, you must add the molar masses of the atoms in $AlBr_3$ and in H_2O to determine the molar mass of each compound.

molar mass $AlBr_3$ = 26.98 g + 3(79.90 g) = 266.68 g/mol

molar mass H_2O = 2(1.008 g) + 15.999 g = 18.015 g/mol

$$\text{moles of AlBr}_3 = \text{grams AlBr}_3 \times \frac{1 \text{ mol AlBr}_3}{\text{grams AlBr}_3}$$

$$\text{moles of AlBr}_3 = 71.16 \text{ g AlBr}_3 \times \frac{1 \text{ mol AlBr}_3}{266.68 \text{ g AlBr}_3} =$$

$$0.2668 \text{ mol AlBr}_3$$

$$\text{moles of H}_2\text{O} = \text{grams H}_2\text{O} \times \frac{1 \text{ mol H}_2\text{O}}{\text{gram H}_2\text{O}}$$

$$\text{moles of H}_2\text{O} = 28.84 \text{ g H}_2\text{O} \times \frac{1 \text{ mol H}_2\text{O}}{18.015 \text{ g H}_2\text{O}} = 1.601 \text{ mol H}_2\text{O}$$

Next, calculate the mole ratio of H_2O to $AlBr_3$.

$$\text{mole ration } \frac{\text{H}_2\text{O}}{\text{AlBr}_3} = \frac{1.601 \text{ mol H}_2\text{O}}{0.2668 \text{ mol AlBr}_3} = 6.001$$

This ratio tells you that the hydrate is made up of six molecules of water for each formula unit of $AlBr_3$. Now you can write the molecular formula for the hydrate as $AlBr_3 \cdot 6H_2O$.

Practice Problems

35. Cerium(III) iodide (CeI_3) occurs as a hydrate with the composition 76.3% CeI_3 and 23.7% H_2O. Calculate the formula for the hydrate.

36. Cobalt(II) nitrate ($Co(NO_3)_2$) is used in ceramic glazes and produces a blue color when the ceramic is fired. This compound exists as a hydrate whose composition is 37.1% water and 62.9% $Co(NO_3)_2$ by mass. Write the molecular formula for this hydrate.

37. Lead(II) acetate ($Pb(C_2H_3O_2)_2$) exists as a crystalline hydrate that is 14.25% water by mass. What is the molecular formula for this hydrate?

38. A 17.44-g sample of a hydrate of zinc sulfate ($ZnSO_4$) is heated strongly in a crucible to drive off all water of hydration. After heating, the sample has a mass of 9.79 g. Calculate the molecular formula for this hydrate. (Hint: The mass lost represents water. First determine the moles of water lost from the hydrate.)

Chapter 11 Review

39. Explain in detail how it is possible to count atoms, ions, formula units, and molecules by measuring mass.

40. Why must a chemist be able to count particles of matter?

41. How is the molar mass of a compound related to the atomic masses of the atoms in the compound?

42. The atomic mass of nitrogen is 14.007 amu. What is the mass in grams of a mole of N_2 molecules?

43. An analysis of silicon dioxide (SiO_2) found in beach sand in California shows that it is 46.74% silicon and 53.26% oxygen. Can you assume that silicon dioxide found in a quartz rock in Michigan will have the same analysis? Explain your answer.

44. The empirical formula for a certain compound is found to be C_2H_4O. From this information, what can you predict about the molecular formula for the compound?

45. The formula for magnesium bromide hexahydrate is $MgBr_2 \cdot 6H_2O$. This compound could be represented by the formula $MgBr_2H_{12}O_6$. Explain why this formula is almost never used.

Stoichiometry

12.1 What is stoichiometry?

Stoichiometry is the study of quantitative relationships between amounts of reactants used and products formed by a chemical reaction. Stoichiometry is based on the law of conservation of mass. The law of conservation of mass states that matter is neither created nor destroyed. Thus, in a chemical reaction, the mass of the reactants equals the mass of the products. You can use stoichiometry to answer questions about the amounts of reactants used or products formed by reactions. For example, look at the balanced chemical equation for the formation of table salt (NaCl).

$$2Na(s) + Cl_2(g) \rightarrow 2NaCl(s)$$

You could use stoichiometry to answer the following questions about the chemical reaction.

- How much sodium is needed to produce a certain amount of table salt?
- How much chlorine is needed to produce a certain amount of table salt?
- Given a certain amount of sodium or chlorine, how much table salt can be produced?

When you look at a balanced equation, there are two ways to interpret the coefficients. The coefficients tell you how many individual particles are interacting in the chemical reaction. For example, from the chemical equation above, you learn that two sodium atoms react with one chlorine molecule to form two formula units of table salt. You also learn that two moles of sodium react with one mole of chlorine to form two moles of salt. What the coefficients do not tell you directly is the masses of the reactants and the products in the chemical reaction.

Example Problem 12-1
Interpreting Chemical Equations

Interpret the balanced chemical equation in terms of particles, moles, and mass. Show that the law of conservation of mass is observed.

$$4NH_3 + 5O_2 \rightarrow 4NO + 6H_2O$$

The coefficients represent both the numbers of particles and the numbers of moles interacting in the chemical reaction.

$$4 \text{ molecules } NH_3 + 5 \text{ molecules } O_2 \rightarrow$$
$$4 \text{ molecules } NO + 6 \text{ molecules } H_2O$$

$$4 \text{ moles } NH_3 + 5 \text{ moles } O_2 \rightarrow 4 \text{ moles } NO + 6 \text{ moles } H_2O$$

You can calculate the mass of each reactant and product by multiplying the number of moles by the conversion factor molar mass.

$$4 \text{ mol } NH_3 \times \frac{17.03 \text{ g } NH_3}{1 \text{ mol } NH_3} = 68.12 \text{ g } NH_3$$

$$5 \text{ mol } O_2 \times \frac{32.00 \text{ g } O_2}{1 \text{ mol } O_2} = 160.0 \text{ g } O_2$$

$$4 \text{ mol } NO \times \frac{30.01 \text{ g } NO}{1 \text{ mol } NO} = 120.0 \text{ g } NO$$

$$6 \text{ mol } H_2O \times \frac{18.02 \text{ g } H_2O}{1 \text{ mol } H_2O} = 108.1 \text{ g } H_2O$$

The law of conservation of mass is observed because the mass of the reactants (68.12 g NH_3 + 160.0 g O_2 = 228.1 g) equals the mass of the products (120.0 g NO + 108.1 g H_2O = 228.1 g).

Practice Problems

1. Interpret each balanced equation in terms of particles, moles, and mass. Show that the law of conservation of mass is observed.

   a. $2H_2O_2(l) \rightarrow O_2(g) + 2H_2O(l)$
   b. $H_2CO_3(aq) \rightarrow H_2O(l) + CO_2(g)$
   c. $4HCl(aq) + O_2(g) \rightarrow 2H_2O(l) + 2Cl_2(g)$

As you know, the coefficients in a balanced chemical equation indicate the relationships among the moles of reactants and products in the reaction. You can use the coefficients to write mole ratios. A **mole ratio** is a ratio between the numbers of moles of any two substances in a balanced chemical equation. What mole ratios can be written for the following chemical equation?

$$2H_2O_2(l) \rightarrow O_2(g) + 2H_2O(l)$$

$$\frac{2 \text{ mol } H_2O_2}{1 \text{ mol } O_2} \text{ and } \frac{2 \text{ mol } H_2O_2}{2 \text{ mol } H_2O}$$

$$\frac{1 \text{ mol } O_2}{2 \text{ mol } H_2O_2} \text{ and } \frac{1 \text{ mol } O_2}{2 \text{ mol } H_2O}$$

$$\frac{2 \text{ mol } H_2O}{2 \text{ mol } H_2O_2} \text{ and } \frac{2 \text{ mol } H_2O}{1 \text{ mol } O_2}$$

To determine the number of mole ratios that defines a given chemical reaction, multiply the number of species in the equation by the next lower number. Thus, a chemical reaction with three participating species can be defined by six mole ratios ($3 \times 2 = 6$); a chemical reaction with four species can be defined by 12 mole ratios ($4 \times 3 = 12$).

Why learn to write mole ratios? They are the key to calculations that are based on chemical equations. Using a balanced chemical equation, mole ratios derived from the equation, and a given amount of one of the reactants or products, you can calculate the amount of any other participant in the reaction.

Practice Problems

2. Determine all the mole ratios for the following balanced chemical equations.
   a. $N_2(g) + O_2(g) \rightarrow 2NO(g)$
   b. $4NH_3(aq) + 5O_2(g) \rightarrow 4NO(g) + 6H_2O(l)$
   c. $4HCl(aq) + O_2(g) \rightarrow 2H_2O(l) + 2Cl_2(g)$

12.2 Stoichiometric Calculations

There are three basic stoichiometric calculations: mole-to-mole conversions, mole-to-mass conversions, and mass-to-mass conversions. All stoichiometric calculations begin with a balanced equation and mole ratios.

▶ **Stoichiometric mole-to-mole conversion** How can you determine the number of moles of table salt (NaCl) produced from 0.02 moles of chlorine (Cl_2)?

First, write the balanced equation.

$$2Na(s) + Cl_2(g) \rightarrow 2NaCl(s)$$

Then, use the mole ratio to convert the known number of moles of chlorine to the number of moles of table salt. Use the formula below.

$$\text{moles of known} \times \frac{\text{moles of unknown}}{\text{moles of known}} = \text{moles of unknown}$$

$$0.02 \text{ mol } Cl_2 \times \frac{2 \text{ mol NaCl}}{1 \text{ mol } Cl_2} = 0.04 \text{ mol } Cl_2$$

Example Problem 12-2
Stoichiometric Mole-to-Mole Conversion

A piece of magnesium burns in the presence of oxygen, forming magnesium oxide (MgO). How many moles of oxygen are needed to produce 12 moles of magnesium oxide?

Write the balanced equation and the mole ratio that relates mol O_2 to mol MgO.

$$2Mg(s) + O_2(g) \rightarrow 2MgO(s)$$

$$\frac{1 \text{ mol } O_2}{2 \text{ mol MgO}}$$

Multiply the known number of moles of MgO by the mole ratio.

$$12 \text{ mol MgO} \times \frac{1 \text{ mol } O_2}{2 \text{ mol MgO}} = 6 \text{ mol } O_2$$

Six moles of oxygen is needed to produce 12 moles of magnesium oxide.

Practice Problems

3. The carbon dioxide exhaled by astronauts can be removed from a spacecraft by reacting it with lithium hydroxide (LiOH). The reaction is as follows: $CO_2(g) + 2LiOH(s) \rightarrow Li_2CO_3(s) + H_2O(l)$. An average person exhales about 20 moles of CO_2 per day. How many moles of LiOH would be required to maintain two astronauts in a spacecraft for three days?

4. Balance the following equation and answer the questions below.

$$KClO_3(s) \rightarrow KCl(s) + O_2(g)$$

a. How many moles of O_2 are produced from 10 moles of $KClO_3$?

b. How many moles of KCl are produced using 3 moles of $KClO_3$?

c. How many moles of $KClO_3$ are needed to produce 50 moles of O_2?

▶ **Stoichiometric mole-to-mass conversion** A mole-to-mass conversion allows you to calculate the mass of a product or reactant in a chemical reaction given the number of moles of a reactant or product.

Example Problem 12-3
Stoichiometric Mole-to-Mass Conversion

The following reaction occurs in plants undergoing photosynthesis.

$$6CO_2(g) + 6H_2O(l) \rightarrow C_6H_{12}O_6(s) + 6O_2(g)$$

How many grams of glucose ($C_6H_{12}O_6$) are produced when 24.0 moles of carbon dioxide reacts in excess water?

Determine the number of moles of glucose produced by the given amount of carbon dioxide.

$$24.0 \text{ mol CO}_2 \times \frac{1 \text{ mol C}_6H_{12}O_6}{6 \text{ mol CO}_2} = 4.00 \text{ mol C}_6H_{12}O_6$$

Multiply by the molar mass.

$$4.00 \text{ mol C}_6H_{12}O_6 \times \frac{180.18 \text{ g C}_6H_{12}O_6}{1 \text{ mol C}_6H_{12}O_6} = 721 \text{ g C}_6H_{12}O_6$$

721 grams of glucose is produced from 24.0 moles of carbon dioxide.

Practice Problems

5. Calculate the mass of sodium chloride (NaCl) produced when 5.50 moles of sodium reacts in excess chlorine gas.

6. How many grams of chlorine gas must be reacted with excess sodium iodide (NaI) to produce 6.00 moles of sodium chloride?

   a. Balance the equation: $NaI(aq) + Cl_2(g) \rightarrow NaCl(aq) + I_2(s)$.

   b. Perform the calculation.

7. Calculate the mass of hydrochloric acid (HCl) needed to react with 5.00 moles of zinc.

   a. Balance the equation: $Zn(s) + HCl(aq) \rightarrow ZnCl_2(aq) + H_2(g)$.

   b. Perform the calculation.

▶ **Stoichiometric mass-to-mass conversion** If you were preparing to carry out a chemical reaction in the laboratory, you would need to know how much of each reactant to use in order to produce a certain mass of product. This is one instance when you would use a mass-to-mass conversion. In this calculation, you can find the mass of an unknown substance in a chemical equation if you have the balanced chemical equation and know the mass of one substance in the equation.

Example Problem 12-4
Stoichiometric Mass-to-Mass Conversion

How many grams of sodium hydroxide (NaOH) are needed to completely react with 50.0 grams of sulfuric acid (H_2SO_4) to form sodium sulfate (Na_2SO_4) and water?

Write the balanced equation.

$$2NaOH(aq) + H_2SO_4(aq) \rightarrow Na_2SO_4 + 2H_2O(g)$$

Convert grams of sulfuric acid to moles NaOH.

$$50.0 \text{ g } H_2SO_4 \times \frac{1 \text{ mol } H_2SO_4}{98.09 \text{ g } H_2SO_4} = 0.510 \text{ mol } H_2SO_4$$

$$0.510 \text{ mol } H_2SO_4 \times \frac{2 \text{ mol NaOH}}{1 \text{ mol } H_2SO_4} = 1.02 \text{ mol NaOH}$$

Calculate the mass of NaOH needed.

$$1.02 \text{ mol NaOH} \times \frac{40.00 \text{ g NaOH}}{1 \text{ mol NaOH}} = 40.8 \text{ g NaOH}$$

50.0 grams of H_2SO_4 reacts completely with 40.8 grams of NaOH.

Practice Problems

8. Balance each equation and solve the problem.

a. If 40.0 g of magnesium reacts with excess hydrochloric acid (HCl), how many grams of magnesium chloride ($MgCl_2$) are produced?

$$Mg(s) + HCl(aq) \rightarrow MgCl_2(aq) + H_2(g)$$

b. Determine the mass of copper needed to react completely with a solution containing 12.0 g of silver nitrate ($AgNO_3$).

$$Cu(s) + AgNO_3(aq) \rightarrow Cu(NO_3)_2(aq) + Ag(s)$$

c. How many grams of hydrogen chloride (HCl) are produced when 15.0 g of sodium chloride (NaCl) reacts with excess sulfuric acid (H_2SO_4)?

$$NaCl(aq) + H_2SO_4(aq) \rightarrow Na_2SO_4 + HCl(g)$$

d. Calculate the mass of silver phosphate (Ag_3PO_4) produced if 30.0 g of silver acetate ($AgCH_3COO$) reacts with excess sodium phosphate (Na_3PO_4).

$$AgCH_3COO(aq) + Na_3PO_4(aq) \rightarrow$$
$$Ag_3PO_4(s) + NaCH_3COO(aq)$$

▶ **Steps in stoichiometric calculations** Follow these basic steps when performing stoichiometric calculations.

1. Write the balanced equation.

2. Determine the moles of the given substance using a mass-to-mole conversion. Use the inverse of the molar mass as the conversion factor.

3. Determine the moles of the unknown substance from the moles of the given substance. Use the mole ratio from the balanced equation as the conversion factor.

4. From the moles of the unknown substance, determine the mass of the unknown substance using a mole-to-mass conversion. Use the molar mass as the conversion factor.

12.3 Limiting Reactants

Rarely are the reactants in a chemical reaction present in the exact mole ratios specified in the balanced equation. Usually, one or more of the reactants are present in excess, and the reaction proceeds until all of one reactant is used up. The reactant that is used up is called the **limiting reactant.** The limiting reactant limits the reaction and, thus, determines how much of the product forms. The left-over reactants are called **excess reactants.**

How can you determine which reactant in a chemical reaction is limited? First, find the number of moles of each reactant by multiplying the given mass of each reactant by the inverse of the molar mass. Next, determine whether the reactants are available in the mole ratio specified in the balanced equation. A reactant that is available in an amount smaller than that required by the mole ratio is a limiting reactant.

After the limiting reactant has been determined, calculate the amount of product that can ideally form from the given amount of the limiting reactant. To do this, multiply the given number of moles of the limiting reactant by the mole ratio that relates the limiting reactant to the product. Then, convert moles of product to mass using the molar mass of the product as the conversion factor.

Example Problem 12-5
Determining the Limiting Reactant

In the reaction below, 40.0 g of sodium hydroxide (NaOH) reacts with 60.0 g of sulfuric acid (H_2SO_4).

$$2NaOH(aq) + H_2SO_4(aq) \rightarrow Na_2SO_4 + 2H_2O(g)$$

a. Which reactant is the limiting reactant?

b. What mass of Na_2SO_4 can be produced using the given quantities of the reactants?

a. To determine the limiting reactant, calculate the actual ratio of available moles of reactants.

$$40.0 \text{ g NaOH} \times \frac{1 \text{ mol NaOH}}{40.0 \text{ g NaOH}} = 1.00 \text{ mol NaOH}$$

$$60.0 \text{ g } H_2SO_4 \times \frac{1 \text{ mol } H_2SO_4}{98.09 \text{ g } H_2SO_4} = 0.612 \text{ mol } H_2SO_4$$

So, $\frac{1.00 \text{ mol NaOH}}{0.612 \text{ mol } H_2SO_4}$ is available. Compare this ratio with the

mole ratio from the balanced equation: $\frac{2 \text{ mol NaOH}}{1 \text{ mol } H_2SO_4}$, or

$\frac{1 \text{ mol NaOH}}{0.5 \text{ mol } H_2SO_4}$. You can see that when 0.5 mol H_2SO_4 has reacted,

all of the 1.00 mol of NaOH would be used up. Some H_2SO_4 would

remain unreacted. Thus, NaOH is the limiting reactant.

b. To calculate the mass of Na_2SO_4 that can form from the given
reactants, multiply the number of moles of the limiting reactant
(NaOH) by the mole ratio of the product to the limiting reactant
and then multiply by the molar mass of the product.

$$1.00 \text{ mol NaOH} \times \frac{1 \text{ mol } Na_2SO_4}{2 \text{ mol NaOH}} \times \frac{142.04 \text{ g } Na_2SO_4}{1 \text{ mol } Na_2SO_4}$$

$$= 71.0 \text{ g } Na_2SO_4$$

71.0 g of Na_2SO_4 can form from the given amounts of the reactants.

Practice Problems

9. Ammonia (NH_3) is one of the most common chemicals produced
in the United States. It is used to make fertilizer and other products. Ammonia is produced by the following chemical reaction.

$$N_2(g) + 3H_2(g) \rightarrow 2NH_3(g)$$

a. If you have 1.00×10^3 g of N_2 and 2.50×10^3 g of H_2,
which is the limiting reactant in the reaction?

b. How many grams of ammonia can be produced from the
amount of limiting reactant available?

c. Calculate the mass of excess reactant that remains after the
reaction is complete.

10. Aluminum reacts with chlorine to produce aluminum chloride.

  a. Balance the equation: $Al(s) + Cl_2(g) \rightarrow AlCl_3(s)$.

  b. If you begin with 3.2 g of aluminum and 5.4 g of chlorine, which is the limiting reactant?

  c. How many grams of aluminum chloride can be produced from the amount of limiting reactant available?

  d. Calculate the mass of excess reactant that remains after the reaction is complete.

Reactions do not always continue until all of the reactants are used up. Using an excess of the least expensive reactant in a reaction can ensure that all of the more expensive reactant is used up, making the chemical reaction more efficient and cost-effective. Using an excess of one reactant can also speed up some reactions.

12.4 Percent Yield

Most chemical reactions do not produce the predicted amount of product. Although your work so far with stoichiometric problems may have led you to believe that chemical reactions proceed according to the balanced equation and always produce the calculated amount of product, it's not true. Many reactions stop before all the reactants are used up, so less product is formed than expected. Also, products other than those expected sometimes form from competing chemical reactions, thereby reducing the amount of the desired product.

The **theoretical yield** is the maximum amount of product that can be produced from a given amount of reactant under ideal circumstances. This is the amount you have been calculating in practice problems so far. Chemical reactions hardly ever produce the theoretical yield. The **actual yield** is the amount of product that is actually produced when a chemical reaction is carried out in an experiment. It is determined by measuring the mass of the product. **Percent yield** of product is the ratio of the actual yield to the theoretical yield expressed as a percent.

$$\text{Percent yield} = \frac{\text{actual yield (from an experiment)}}{\text{theoretical yield (from stoichiometric calculations)}} \times 100$$

Percent yield tells you how efficient a chemical reaction is in producing the desired product.

Example Problem 12-6
Calculating Percent Yield

Aspirin ($C_9H_8O_4$) can be made from salicylic acid ($C_7H_6O_3$) and acetic anhydride ($C_4H_6O_3$). Suppose you mix 13.2 g of salicylic acid with an excess of acetic anhydride and obtain 5.9 g of aspirin and some water. Calculate the percent yield of aspirin in this reaction.

Write the balanced equation.

$$2C_7H_6O_3(s) + C_4H_6O_3(l) \rightarrow 2C_9H_8O_4(s) + H_2O(l)$$

Calculate the theoretical yield. Salicylic acid is the limiting reactant.

$$13.2 \text{ g } C_7H_6O_3 \times \frac{1 \text{ mol } C_7H_6O_3}{138.1 \text{ g } C_7H_6O_3} = 0.0956 \text{ mol } C_7H_6O_3$$

$$0.0956 \text{ mol } C_7H_6O_3 \times \frac{2 \text{ mol } C_9H_8O_4}{2 \text{ mol } C_7H_6O_3} = 0.0956 \text{ mol } C_9H_8O_4$$

$$0.0956 \text{ mol } C_9H_8O_4 \times \frac{180.2 \text{ g } C_9H_8O_4}{1 \text{ mol } C_9H_8O_4} = 17.2 \text{ g } C_9H_8O_4$$

Calculate the percent yield.

$$\frac{5.9 \text{ g } C_9H_8O_4}{17.2 \text{ g } C_9H_8O_4} \times 100 = 34\%$$

Practice Problems

11. Calculate the percent yield for each chemical reaction based on the data provided.

    a. theoretical yield: 25 g; actual yield: 20 g

    b. theoretical yield: 55 g; actual yield: 42 g

    c. theoretical yield: 5.2 g; actual yield: 4.9 g

12. Calculate the actual yield for each chemical reaction based on the data provided.

    a. theoretical yield: 20 g; percent yield: 95%

    b. theoretical yield: 75 g; percent yield: 88%

    c. theoretical yield: 9.2 g; percent yield: 62%

13. In an experiment, 10.0 g of magnesium reacted with excess hydrochloric acid forming magnesium chloride.

$$Mg(s) + 2HCl(aq) \rightarrow MgCl_2(aq) + H_2(g)$$

At the completion of the reaction, 29.5 g of magnesium chloride was produced. Calculate the theoretical yield and the percent yield.

Percent yield is important in the calculation of overall cost effectiveness in industrial processes. Manufacturers must reduce the cost of making products to the lowest level possible. For example, sulfuric acid (H_2SO_4) is a raw material for many products, including fertilizers, detergents, pigments, and textiles. The cost of sulfuric acid affects the cost of many consumer items that use sulfuric acid as a raw material.

The manufacture of sulfuric acid is sometimes achieved using a two-step process called the contact process. The two steps are as follows.

$$S_8(s) + 8O_2(g) \rightarrow 8SO_2(g)$$

$$2SO_2(g) + O_2(g) \rightarrow 2SO_3(g)$$

The last step results in sulfuric acid as the product.

$$SO_3(g) + H_2O(l) \rightarrow H_2SO_4(aq)$$

The first step produces almost 100 percent yield. The second step will also produce a high yield if a catalyst is used. A catalyst is a substance that speeds up a chemical reaction but is not used up in the chemical reaction and does not appear in the chemical equation.

Chapter 12 Review

14. What is stoichiometry?

15. Write two questions that stoichiometry can help you answer about the following chemical equation.

$$4HCl(aq) + O_2(g) \rightarrow 2H_2O(l) + 2Cl_2(g)$$

16. Relate the law of conservation of mass to stoichiometry.

17. What is the difference between a limiting reactant and an excess reactant?

States of Matter

13.1 Gases

In the late 1800s, two scientists, Ludwig Boltzmann and James Maxwell, independently proposed a model to explain the properties of gases in terms of particles in motion. This model is now known as the **kinetic-molecular theory.** The model makes the following assumptions about the size, motion, and energy of gas particles.

- **Particle size** The particles in a gas are separated from one another by empty space. The volume of the empty space is much greater than the volume of the gas particles themselves. Because gas particles are so far apart, there are no significant attractive or repulsive forces between them.

- **Particle motion** Gas particles are in constant, random motion. Until they bump into something (another particle or the side of a container), particles move in a straight line. When gas particles do collide with something, the collision is said to be elastic. An **elastic collision** is one in which no kinetic energy is lost. Although kinetic energy may be transferred from one particle to another, the total amount of kinetic energy of the two particles does not change.

- **Particle energy** Mass and velocity determine the kinetic energy of a particle, as represented in the equation below.

$$KE = \frac{1}{2} mv^2$$

KE = kinetic energy
m = mass of the particle
v = velocity of the particle

The velocity of a particle includes both its speed and its direction. Each particle in a sample containing only one gas will have the same mass but not the same velocity. Thus, all the particles in a sample of gas do not have the same kinetic energy. **Temperature** is a measure of the *average* kinetic energy of the particles in a sample of matter. At a given temperature, all gases have the same average kinetic energy.

▶ **Explaining the behavior of gases** The kinetic-molecular theory explains the following behavior of gases.

- **Low density** Density is a measure of mass per unit volume. The difference between the high density of a solid and the low density of a gas is due mainly to the large amount of space between the particles in the gas. There are fewer particles in a gas than in a solid of the same volume.

- **Compression and expansion** A gas will expand to fill its container. Thus, the density of a sample of gas will change with the volume of the container it is placed in. The gas will become more dense as it is compressed into a smaller container. The gas will become less dense as it expands in a larger container.

- **Diffusion** Gas particles flow past each other easily because there are no significant forces of attraction between them. **Diffusion** refers to the movement of one material through another, such as when one gas flows into a space already occupied by another gas. The rate of diffusion depends mostly on the mass of the particles. Lighter particles diffuse more quickly than heavier particles. Because lighter particles have the same average kinetic energy as do heavier particles at the same temperature, lighter particles must have, on average, a greater velocity.

- **Effusion** If you have ever seen a tire deflate from a puncture, you are familiar with effusion. Effusion is the escape of a gas through a small opening in its container. **Graham's law of effusion** states that the rate of effusion for a gas is inversely proportional to the square root of its molar mass.

$$\text{Rate of effusion} \propto \frac{1}{\sqrt{\text{molar mass}}}$$

Using Graham's law, you can also compare the rates of diffusion for two gases.

$$\frac{\text{Rate}_A}{\text{Rate}_B} = \sqrt{\frac{\text{molar mass}_B}{\text{molar mass}_A}}$$

Example Problem 13-1
Finding the Ratio of Diffusion Rates

The molar mass of helium is 4.00 g/mol; the molar mass of air is
29.0 g/mol. What is the ratio of their diffusion rates? Which gas
diffuses faster?

$$\frac{\text{Rate}_{He}}{\text{Rate}_{Air}} = \sqrt{\frac{\text{molar mass}_{Air}}{\text{molar mass}_{He}}}$$

$$= \sqrt{\frac{29.0 \text{ g/mol}}{4.00 \text{ g/mol}}} = \sqrt{7.25} = 2.69$$

The ratio of the diffusion rates is 2.69. Helium diffuses about
2.7 times faster than air does.

Practice Problems

1. Calculate the ratio of diffusion rates for neon and helium.
   Which gas diffuses faster? About how much faster?
2. Calculate the ratio of diffusion rates for ammonia (NH_3) and
   carbon dioxide (CO_2). Which gas diffuses more rapidly?
3. What is the ratio of diffusion rates for argon and radon? Which
   gas diffuses more rapidly?
4. Tetrafluoroethylene (C_2F_4) effuses at a rate of 4.6×10^{-6}
   moles per hour. An unknown gas effuses at a rate of 6.5×10^{-6}
   moles per hour under identical conditions. What is the molar
   mass of the unknown gas?

▶ **Gas pressure** When gas particles collide with the walls of their
container, they exert pressure on the walls. **Pressure** is force per unit
area. The pressure exerted by the particles in the atmosphere that
surrounds Earth is called atmospheric pressure, or air pressure. Air
pressure varies at different locations on Earth. At Earth's surface,
air pressure is approximately equal to the pressure exerted by a
1-kilogram mass on a square centimeter. Air pressure at higher alti-
tudes, such as on a mountaintop, is slightly lower than air pressure
at sea level.

Air pressure is measured using a **barometer.** A barometer consists of a thin tube closed on one end and filled with mercury. The

tube is placed so that the level of the mercury is determined by air pressure. The mercury rises when the air pressure increases and falls when the air pressure decreases.

A manometer is an instrument used to measure gas pressure in a closed container. A flask containing gas is attached to a sealed U-shaped tube that contains mercury. The mercury is level across the two arms of the U. When a valve between the flask and the tube is opened, gas particles diffuse into the tube and push down on the mercury. The difference in the height of the mercury in the two arms of the U is used to calculate the pressure of the gas in the flask.

The pascal (Pa) is the SI unit of pressure. One **pascal** is equal to a force of one newton per square meter. Some scientists use other units of pressure. For example, engineers use pounds per square inch. Barometers and manometers measure pressure in millimeters of mercury (mm Hg). A unit called the torr is equal to 1 mm Hg.

Air pressure is often reported in a unit called an atmosphere (atm). One **atmosphere** is equal to 760 mm Hg, 760 torr, or 101.3 kilopascals (kPa). These are all defined units; therefore, they have as many significant figures as needed when used in calculations.

▶ **Dalton's law of partial pressures** Dalton found that each gas in a mixture exerts pressure independently of the other gases. **Dalton's law of partial pressures** states that the total pressure of a mixture of gases is equal to the sum of the pressures of all the gases in the mixture, as shown below.

$$P_{total} = P_1 + P_2 + P_3 + \ldots P_n$$

The portion of the total pressure (P_{total}) exerted by one of the gases is called its partial pressure (P_n). The partial pressure of a gas depends on the number of moles of the gas, the size of the container, and the temperature of the mixture. The partial pressure of one mole of any gas is the same at a given temperature and pressure.

Example Problem 13-2
Finding the Partial Pressure of a Gas

Air is made up of four main gases: N_2, O_2, Ar, and CO_2. Air pressure at sea level is approximately 760 mm Hg. Calculate the partial pressure of oxygen, given the following partial pressures: N_2, 594 mm Hg; Ar, 7.10 mm Hg; and CO_2, 0.27 mm Hg.

Use Dalton's law of partial pressures to solve the problem.

$$P_{total} = P_{N_2} + P_{Ar} + P_{CO_2} + P_{O_2}$$
$$P_{O_2} = P_{total} - P_{N_2} - P_{Ar} - P_{CO_2}$$
$$P_{O_2} = 760 \text{ mm Hg} - 594 \text{ mm Hg} - 7.10 \text{ mm Hg} - 0.27 \text{ mm Hg}$$
$$P_{O_2} = 158.63 \text{ mm Hg}$$

The partial pressure of oxygen is about 159 mm Hg.

Practice Problems

5. What is the partial pressure of oxygen gas in a mixture of nitrogen gas and oxygen gas with a total pressure of 0.48 atm if the partial pressure of nitrogen gas is 0.24 atm?

6. Find the total pressure of a mixture that contains three gases with the following partial pressures: 6.6 kPa, 3.2 kPa, and 1.2 kPa.

7. Find the total pressure of a mixture that contains five gases with the following partial pressures: 7.81 kPa, 13.20 kPa, 2.43 kPa, 12.50 kPa, and 2500 Pa.

8. Find the partial pressure of ammonia in a mixture of three gases with a total pressure of 75.6 kPa if the sum of the partial pressures of the other two gases is 34.9 kPa.

You can use Dalton's law of partial pressures to determine the amount of gas produced by a reaction. Collect the gas in an inverted container of water. The gas will displace the water. If the gas does not react with the water, the total pressure inside the container will be the sum of the partial pressures of the gas and water vapor. The partial pressure of water vapor has a fixed value at a given temperature, which you can look up in a reference chart. You can then calculate the partial pressure of the gas by subtracting the partial pressure of water vapor at that temperature from the total pressure.

13.2 Forces of Attraction

The attractive forces that hold particles together in ionic, covalent, and metallic bonds are called intramolecular forces. Intermolecular forces, which are weaker than intramolecular forces, also can hold particles together. Three types of intermolecular forces are described below: dispersion forces, dipole–dipole forces, and hydrogen bonds.

▶ **Dispersion forces** Weak forces that result from temporary shifts in the density of electrons in electron clouds are called **dispersion forces,** or London forces. When two nonpolar molecules are in close contact, the electron cloud of one molecule repels the electron cloud of the other molecule. As a result, the electron density in each electron cloud is greater in one region of the cloud. Two temporary dipoles form. Weak dispersion forces exist between oppositely charged regions of the dipoles. Dispersion forces, which are the weakest intermolecular forces, are important only when no stronger forces are acting on the particles. Dispersion forces are noticeable between identical nonpolar molecules as the number of electrons involved increases. For example, an increase in dispersion forces explains why fluorine and chlorine are gases, bromine is a liquid, and iodine is a solid at room temperature.

▶ **Dipole–dipole forces** Attractions between oppositely charged regions of polar molecules are called **dipole–dipole forces.** Polar molecules have a permanent dipole and orient themselves so that oppositely charged regions match up. Dipole–dipole forces are stronger than dispersion forces as long as the molecules being compared are similar in mass.

▶ **Hydrogen bonds** A **hydrogen bond** is a dipole–dipole attraction that occurs between molecules containing a hydrogen atom bonded to a small, highly electronegative atom with at least one lone electron pair. The hydrogen must be bonded to a fluorine, an oxygen, or a nitrogen atom. Hydrogen bonds explain why water is a liquid at room temperature, while compounds of comparable mass are gases.

13.3 Liquids and Solids

The kinetic-molecular theory also explains the behavior of liquids and solids. However, the forces of attraction between particles in liquids and solids must be considered as well as their energy of motion.

▶ **Liquids** Liquids conform to the shape of their container but have a fixed volume. The particles in a liquid maintain a fixed volume because the forces of attraction between them limit their range of motion.

- **Density and compression** The density of a liquid is much greater than that of its vapor at the same conditions. The higher density is due to intermolecular forces, which hold the particles together. Although liquids can be compressed, their compression requires an enormous amount of pressure and results in a relatively small change in volume.

- **Fluidity** Fluidity is the ability to flow. Liquids are less fluid than gases.

- **Viscosity** A measure of the resistance of a liquid to flow is called **viscosity.** The stronger the intermolecular forces, the higher is the viscosity. Viscosity also increases with the mass of a liquid's particles and the length of molecule chains. Viscosity decreases with temperature because the average kinetic energy of the particles increases. The added energy makes it easier for the particles to overcome the intermolecular forces that keep the particles from flowing.

- **Surface tension** The energy required to increase the surface area of a liquid by a given amount is called **surface tension.** Surface tension is a measure of the inward pull by particles in the interior of the liquid. The stronger the attractions between particles, the greater is the surface tension. Multiple hydrogen bonds give water a high surface tension.

- **Capillary action** The movement of a liquid up a narrow glass tube is called capillary action, or capillarity. Capillary action occurs when adhesive forces are greater than cohesive forces. Adhesion is the force of attraction between molecules that are different, such as water molecules and the molecules of silicon dioxide in glass. Cohesion is the force of attraction between identical molecules, such as water molecules.

▶ **Solids** Strong attractive forces between the particles in a solid limit the movement of the particles to vibrations around fixed locations. Thus, solids have a definite shape and volume.

Generally, the particles in a solid are more closely packed than those in a liquid, making most solids more dense than most liquids. When the solid and liquid states of a substance coexist, usually the liquid is less dense than the solid. Liquid water and ice are an exception to this rule. Because solids are so dense, ordinary amounts of pressure will not compress them into a smaller volume.

▶ **Crystalline solids** A solid whose atoms, ions, or molecules are arranged in an orderly, geometric, three-dimensional structure (lattice) is called a **crystalline solid.** The individual pieces of a crystalline solid are called crystals. Crystalline solids are divided into five categories based on the types of particles they contain: atomic solids, molecular solids, covalent network solids, ionic solids, and metallic solids. Noble gases are atomic solids whose properties reflect the weak dispersion forces between the atoms.

Molecular solids are held together by dispersion forces, dipole–dipole forces, or hydrogen bonds. Most molecular compounds that are solid at room temperature, such as sugar, have a large molar mass and are poor conductors of heat and electricity.

Elements that are able to form multiple covalent bonds, such as carbon and silicon, are able to form covalent network solids.

The type of ions and the ratio of ions determine the structure of the lattice and the shape of the crystal in an ionic solid. The network of attractions gives these solids high melting points and hardness. They are strong, but brittle, and will shatter when struck.

Metallic solids consist of positive metal ions surrounded by a sea of mobile electrons. The mobile electrons make these solids good conductors of heat and electricity.

Not all solids are crystalline. The particles in an **amorphous solid** are not arranged in a regular, repeating pattern and do not form crystals. Examples of amorphous solids include glass, rubber, and many plastics.

Practice Problems

9. Classify each crystalline solid as molecular, ionic, covalent network, or metallic.

a. NaCl   **c.** Fe

b. SiO_2 **d.** H_2O

13.4 Phase Changes

Most substances can exist in three states—solid, liquid, and gas—depending on the temperature and pressure. States of substances are called phases when they coexist as physically distinct parts of a mixture, such as ice water. When energy is added to or taken away from a system, one phase can change into another.

▶ **Phase changes that require energy** You know the three phases of water: ice, liquid water, and water vapor. When you add ice to water, heat flows from the water to the ice and disrupts the hydrogen bonds that hold the water molecules in the ice together. The ice melts and becomes liquid. The amount of energy required to melt one mole of a solid depends on the strength of the forces keeping the particles together. The **melting point** of a crystalline solid is the temperature at which the forces holding the crystal lattice together are broken and the solid becomes a liquid. Because amorphous solids tend to act like liquids when they are in the solid state, it's hard to specify their melting points.

When liquid water is heated, some molecules escape from the liquid and enter the gas phase. If a substance is usually a liquid at room temperature (as water is), the gas phase is called a vapor. **Vaporization** is the process by which a liquid changes into a gas or vapor. When vaporization occurs only at the surface of a liquid, the process is called **evaporation.**

Vapor pressure is the pressure exerted by a vapor over a liquid. As temperature increases, water molecules gain kinetic energy and vapor pressure increases. When the vapor pressure of a liquid equals atmospheric pressure, the liquid has reached its **boiling point,** which is 100°C for water at sea level. At this point, molecules throughout the liquid have the energy to enter the gas or vapor phase.

The process by which a solid changes directly into a gas without first becoming a liquid is called **sublimation.** Solid air fresheners and dry ice are examples of solids that sublime. At very low temperatures, ice will sublime in a short amount of time. This property of ice is used to preserve freeze-dried foods.

Practice Problems

10. Classify each of the following phase changes.

   a. dry ice (solid carbon dioxide) to carbon dioxide gas

   b. ice to liquid water

   c. liquid bromine to bromine vapor

   d. moth balls giving off a pungent odor

▶ **Phase changes that release energy** Some phase changes release energy into their surroundings. For example, when a vapor loses energy, it may change into a liquid. **Condensation** is the process by which a gas or vapor becomes a liquid. It is the reverse of vaporization. Water vapor undergoes condensation when its molecules lose energy, their velocity decreases, and hydrogen bonds begin to form between them. When hydrogen bonds form, energy is released.

When water is placed in a freezer, heat is removed from the water. The water molecules lose kinetic energy, and their velocity decreases. When enough energy has been removed, the hydrogen bonds keep the molecules frozen in set positions. The **freezing point** is the temperature at which a liquid becomes a crystalline solid.

When a substance changes from a gas or vapor directly into a solid without first becoming a liquid, the process is called **deposition.** Deposition is the reverse of sublimation. Frost is an example of water deposition.

Practice Problems

11. Classify each of the following phase changes.

   a. liquid water to ice

   b. water vapor to liquid water

   c. dew forming on grass

   d. water vapor to ice crystals

   e. beads of water forming on the outside of a glass containing a cold drink

▶ **Phase diagrams** Temperature and pressure control the phase of a substance. A **phase diagram** is a graph of pressure versus temperature that shows in which phase a substance exists under different conditions of temperature and pressure. A phase diagram typically has three regions, each representing a different phase and three curves that separate each phase. The points on the curves indicate conditions under which two phases coexist. The phase diagram for each substance is different because the normal boiling and freezing points of substances are different.

The **triple point** is the point on a phase diagram that represents the temperature and pressure at which three phases of a substance can coexist. All six phase changes can occur at the triple point: freezing and melting, evaporation and condensation, sublimation and deposition. The critical point indicates the critical pressure and the critical temperature above which a substance cannot exist as a liquid.

Practice Problems

12. Answer the following questions about the phase diagram for water in Figure 13-28a on page 409 of your textbook.

    a. List the phase changes a sample of ice would go through if heated to its critical temperature at 1 atm pressure.

    b. At what range of pressure will water be a liquid at temperatures above its normal boiling point?

    c. In what phase does water exist at its triple point?

13. Answer the following questions about the phase diagram for carbon dioxide in Figure 13-28b on page 409 of your textbook.

    a. What happens to solid carbon dioxide at room temperature at 1 atm pressure?

    b. The triple point for carbon dioxide is 5 atm and $-57°C$. List the phase changes a sample of dry ice would go through if heated from $-100°C$ to $10°C$ at 6 atm pressure.

Chapter 13 Review

14. List three assumptions of the kinetic-molecular theory.

15. Which has a lower density—oxygen gas or silver? Explain your answer.

16. Distinguish between diffusion and effusion.

17. Which of these gases will diffuse the fastest: CO, Xe , or He? The slowest? Explain your answers.

18. Do all of the particles in a sample of nitrogen gas have the same kinetic energy at 100 K? Explain your answer.

19. Use what you learned in this chapter about viscosity to explain which container of maple syrup will take the longest to pour over pancakes: a container that has been refrigerated, one that has been microwaved, or one kept at room temperature?

20. List three properties of liquids.

21. Explain why some liquids form a convex surface when placed in a narrow test tube, while others form a concave surface.

22. Why do gases diffuse more quickly than do liquids?

23. What general property of solids accounts for their high density?

24. How does an amorphous solid differ from a crystalline solid?

25. Identify each process.

a. solid → liquid

b. solid → gas

c. liquid → solid

d. liquid → gas

e. gas → liquid

f. gas → solid

Gases

14.1 The Gas Laws

The gas laws apply to ideal gases, which are described by the kinetic theory in the following five statements.

- Gas particles do not attract or repel each other.
- Gas particles are much smaller than the spaces between them.
- Gas particles are in constant, random motion.
- No kinetic energy is lost when gas particles collide with each other or with the walls of their container.
- All gases have the same kinetic energy at a given temperature.

▶ **Boyle's Law** At a constant temperature, the pressure exerted by a gas depends on the frequency of collisions between gas particles and the container. If the same number of particles is squeezed into a smaller space, the frequency of collisions increases, thereby increasing the pressure. Thus, **Boyle's law** states that at constant temperature, the pressure and volume of a gas are inversely related. In mathematical terms, this law is expressed as follows.

$$P_1V_1 = P_2V_2$$

Example Problem 14-1
Applying Boyle's Law

A sample of compressed methane has a volume of 648 mL at a pressure of 503 kPa. To what pressure would the methane have to be compressed in order to have a volume of 216 mL?

Examine the Boyle's law equation. You need to find P_2, the new pressure, so solve the equation for P_2.

$$P_2 = P_1 \left(\frac{V_1}{V_2} \right)$$

Substitute known values and solve.

$$P_2 = 503 \text{ kPa} \left(\frac{648 \text{ mL}}{216 \text{ mL}} \right)$$
$$P_2 = 1510 \text{ kPa}$$

To determine whether your answer is reasonable, notice that the gas is being squeezed to a smaller volume, which requires that the pressure be increased. So, your answer is probably correct.

Practice Problems

1. What pressure will be needed to reduce the volume of 77.4 L of helium at 98.0 kPa to a volume of 60.0 L?

2. A 250.0-mL sample of chlorine gas is collected when the barometric pressure is 105.2 kPa. What is the volume of the sample after the barometer drops to 100.3 kPa?

3. A weather balloon contains 59.5 L of helium at sea level, where the atmospheric pressure is 101.3 kPa. The balloon is released from a 4000-m mountaintop where the pressure is 61.7 kPa. What is the volume of the balloon when it is released?

4. Meteorologists want the balloon in problem 3 to float at an altitude of 35 000 m where the pressure is 575 Pa. What volume will the balloon have at that altitude?

▶ **Charles's Law** When the temperature of a sample of gas is increased and the volume is free to change, the pressure of the gas does not increase. Instead, the volume of the gas increases in proportion to the increase in Kelvin temperature. This observation is **Charles's law**, which can be stated mathematically as follows.

$$\frac{V_1}{T_1} = \frac{V_2}{T_2}$$

Example Problem 14-2
Applying Charles's Law

A weather balloon contains 5.30 kL of helium gas when the temperature is 12 °C. At what temperature will the balloon's volume have increased to 6.00 kL?

Start by converting the given temperature to kelvins.

$$T_1 = (12 + 273) \text{ K} = 285 \text{ K}$$

Next, solve the Charles's law equation for the new temperature, T_2.

$$\frac{V_1}{T_1} = \frac{V_2}{T_2}; \qquad V_1T_2 = V_2T_1; \qquad T_2 = T_1\left(\frac{V_2}{V_1}\right)$$

Then, substitute the known values and compute the result.

$$T_2 = T_1\left(\frac{V_2}{V_1}\right) = 285 \text{ K}\left(\frac{6.00 \text{ kL}}{5.30 \text{ kL}}\right) = 323 \text{ K}$$

Finally, convert the Kelvin temperature back to Celsius.

$$\text{New temperature} = 323 - 273 = 50°\text{C}$$

To check your answer, take note of the conditions in the problem. The volume of the gas is increasing, so you can predict that the new temperature will be higher than the original temperature.

Practice Problems

5. A sample of SO_2 gas has a volume of 1.16 L at a temperature of 23°C. At what temperature will the gas have a volume of 1.25 L?

6. A balloon is inflated with 6.22 L of helium at a temperature of 36°C. What is the volume of the balloon when the temperature is 22°C?

7. A student collects a 125.0-mL sample of hydrogen. Later, the sample is found to have a volume of 128.6 mL at a temperature of 26°C. At what temperature was the hydrogen collected?

8. A balloon has a volume of 10 500 L if the temperature is 15°C. If the temperature is −35°C, what will be the volume of the balloon?

14.2 The Combined Gas Law and Avogadro's Principle

The gas laws may be combined into a single law, called the **combined gas law**, that relates two sets of conditions of pressure, volume, and temperature by the following equation.

$$\frac{P_1V_1}{T_1} = \frac{P_2V_2}{T_2}$$

With this equation, you can find the value of any one of the variables if you know the other five.

Example Problem 14-3
Applying the Combined Gas Law

A sample of nitrogen monoxide has a volume of 72.6 mL at a temperature of 16°C and a pressure of 104.1 kPa. What volume will the sample occupy at 24°C and 99.3 kPa?

Start by converting the temperatures to kelvins.

$$16°C = (16 + 273) \text{ K} = 289 \text{ K}; 24°C = (24 + 273) \text{ K} = 297 \text{ K}$$

Next, solve the combined gas law equation for the quantity to be determined, the new volume, V_2.

$$\frac{P_1 V_1}{T_1} = \frac{P_2 V_2}{T_2}; \qquad P_1 V_1 T_2 = P_2 V_2 T_1; \qquad V_2 = \frac{P_1 V_1 T_2}{P_2 T_1}$$

Substitute the known quantities and compute V_2.

$$V_2 = \frac{P_1 V_1 T_2}{P_2 T_1} = \frac{104.1 \text{ kPa} \times 72.6 \text{ mL} \times 297 \text{ K}}{99.3 \text{ kPa} \times 289 \text{ K}} = 78.2 \text{ mL}$$

Finally, determine whether the answer makes sense. The temperature increases, which should cause the gas to expand. The decrease in pressure should cause the volume to expand further. Therefore, the final volume should be somewhat larger than the starting volume.

Practice Problems

9. A sample of ammonia gas occupies a volume of 1.58 L at 22°C and a pressure of 0.983 atm. What volume will the sample occupy at 1.00 atm and 0°C?

10. A student collects 285 mL of O_2 gas at a temperature of 15°C and a pressure of 99.3 kPa. The next day, the same sample occupies 292 mL at a temperature of 11°C. What is the new pressure of the gas?

11. A balloon is inflated with 2.42 L of helium at a temperature of 27°C. Later, the volume of the balloon has changed to 2.37 L at a temperature of 19°C and a pressure of 99.7 kPa. What was the pressure when the balloon was inflated?

▶ **Avogadro's Principle** In the early nineteenth century, Avogadro proposed the idea that equal volumes of all gases at the same conditions of temperature and pressure contain the same number of particles. An extension of **Avogadro's principle** is that one mole (6.02×10^{23} particles) of any gas at standard temperature and pressure (0°C and 1.00 atm pressure, STP) occupies a volume of 22.4 L. Given that the mass of a mole of any gas is the molecular mass of the gas expressed in grams, Avogadro's principle allows you to interrelate mass, moles, pressure, volume, and temperature for any sample of gas.

Example Problem 14-4
Applying Avogadro's Principle

What is the volume of 7.17 g of neon gas at 24°C and 1.05 atm? Start by converting the mass of neon to moles. The periodic table tells you that the atomic mass of neon is 20.18 amu. Therefore, the molar mass of neon is 20.18 g.

$$\text{moles Ne} = 7.17 \text{ g Ne} \times \frac{1 \text{ mol Ne}}{20.18 \text{ g Ne}} = 0.355 \text{ mol Ne}$$

Next, determine the volume at STP of 0.355 mol Ne.

$$V_{Ne} = 0.355 \text{ mol Ne} \times \frac{22.4 \text{ L Ne}}{1 \text{ mol Ne}} = 7.95 \text{ L Ne}$$

If you needed only the volume at STP, you could stop here.

Finally, use the combined gas law equation to determine the volume of the neon at 24°C and 1.05 atm pressure.

$$\frac{P_1 V_1}{T_1} = \frac{P_2 V_2}{T_2}; \qquad P_1 V_1 T_2 = P_2 V_2 T_1; \qquad V_2 = \frac{P_1 V_1 T_2}{P_2 T_1}$$

$$V_2 = \frac{P_1 V_1 T_2}{P_2 T_1} = \frac{1.00 \text{ atm} \times 7.95 \text{ L Ne} \times 297 \text{ K}}{1.05 \text{ atm} \times 273 \text{ K}} = 8.24 \text{ L Ne}$$

Practice Problems

12. How many moles of acetylene (C_2H_2) gas occupy a volume of 3.25 L at STP?

13. Determine the volume of 12.3 g of formaldehyde gas (CH_2O) at STP.

14. What is the volume of 1.000 kg of helium gas at 36°C and a pressure of 98.7 kPa?

15. What is the mass of 187 L of CO_2 gas? The volume is measured at STP. (Hint: Start by determining the number of moles of gas present.)

16. How many moles of silane gas (SiH_4) are present in 8.68 mL measured at 18°C and 1.50 atm? (Hint: First determine the volume of the silane at STP.)

14.3 The Ideal Gas Law

The pressure, volume, temperature, and number of moles of gas can be related in a simpler, more convenient way by using the ideal gas law. The following is the law's mathematical expression,

$$PV = nRT,$$

where n represents the number of moles. The **ideal gas constant,** R, already contains the molar volume of a gas at STP along with the standard temperature and pressure conditions. The constant R does the job of correcting conditions to STP. You do not have to correct to STP in a separate step as in Example Problem 14-4. The value of R depends on the units in which the pressure of the gas is measured, as shown below.

$$R = 8.314 \frac{L \cdot kPa}{mol \cdot K} = 0.0821 \frac{L \cdot atm}{mol \cdot K} = 62.4 \frac{L \cdot mm\ Hg}{mol \cdot K}$$

These values are all equivalent. Use the one that matches the pressure units you are using.

Example Problem 14-5
Applying the Ideal Gas Law 2

What pressure in atmospheres will 18.6 mol of methane exert when it is compressed in a 12.00-L tank at a temperature of 45°C?

As always, change the temperature to kelvins before doing anything else.

$$45°C = (45 + 273)\ K = 318\ K$$

Next, solve the ideal gas law equation for P.

$$PV = nRT; \qquad P = \frac{nRT}{V}$$

Substitute the known quantities and calculate P.

$$P = \frac{nRT}{V} = \frac{18.6 \text{ mol} \times 0.0821 \frac{L \cdot atm}{mol \cdot K} \times 318 \, K}{12.00 \, L} = 40.5 \text{ atm}$$

Notice that this pressure makes sense because a large amount of gas is being squeezed into a much smaller space.

Practice Problems

17. What is the pressure in atmospheres of 10.5 mol of acetylene in a 55.0-L cylinder at 37°C?

18. What volume does 0.056 mol of H_2 gas occupy at 25°C and 1.11 atm pressure?

19. A sample of carbon monoxide has a volume of 344 mL at 85°C and a pressure of 88.4 kPa. Determine the amount in moles of CO present.

▶ **Using Mass with the Ideal Gas Law** Recall that it is possible to calculate the number of moles of a sample of a substance when you know the mass of the sample and the formula of the substance.

$$\text{number of moles } (n) = \frac{\text{mass of substance in grams}}{\text{molar mass of substance}} ; \ n = \frac{m}{M}$$

You can substitute this expression into the ideal gas law equation in place of n.

$$PV = nRT; \quad PV = \left(\frac{m}{M}\right)RT; \text{ or } PVM = mRT$$

Notice that this equation enables you to determine the molar mass of a substance if you know the values of the other four variables.

Example Problem 14-6

Determining Molar Mass

Determine the molar mass of an unknown gas if a sample has a mass of 0.290 g and occupies a volume of 148 mL at 13°C and a pressure of 107.0 kPa.

First, convert the temperature to kelvins.

$$13°C = (13 + 273) \, K = 286 \, K$$

Next, solve the ideal gas law equation for M, the molar mass.

$$PVM = mRT; M = \frac{mRT}{PV}$$

Finally, substitute values and calculate the value of M. Notice that you must use the value of R that uses kilopascals as pressure units and express the volume in liters.

$$M = \frac{mRT}{PV} = \frac{0.290 \text{ g} \times 8.314 \frac{L \cdot kPa}{mol \cdot K} \times 286 \text{ } K}{107.0 \text{ } kPa \times 0.148 \text{ } L} = 43.5 \text{ } \frac{g}{mol}$$

Notice that the units cancel to leave grams per mole, the appropriate units for molar mass.

Practice Problems

20. A 250.0-mL sample of a noble gas collected at 88.1 kPa and 7°C has a mass of 0.378 g. What is the molar mass of the gas? Identify the sample.

21. A sample of gas is known to be either H_2S or SO_2. A 2.00-g sample of the gas occupies a volume of 725 mL at a temperature of 13°C and a pressure of 102.4 kPa. What are the molar mass and the identity of the gas?

22. What volume is occupied by 1.000 g of H_2O vapor at a temperature of 134°C and a pressure of 0.0552 atm?

23. A 5.25-L tank contains 87.0 g of neon gas. At what temperature will the tank have a pressure of 19.0 atm?

14.4 Gas Stoichiometry

In Chapter 13, you learned how to use moles and molar mass along with a balanced chemical equation to calculate the masses of reactants and products in a chemical reaction. Now that you know how to relate volumes, masses, and moles for a gas, you can do stoichiometric calculations for reactions involving gases.

Example Problem 14-7
Gas Stoichiometry Using Mass

Ammonium sulfate can be prepared by a reaction between ammonia gas and sulfuric acid as follows.

$$2NH_3(g) + H_2SO_4(aq) \rightarrow (NH_4)_2SO_4(aq)$$

What volume of NH_3 gas, measured at 78°C and a pressure of 1.66 atm, will be needed to produce 5.00×10^3 g of $(NH_4)_2SO_4$?
First, you need to compute the number of moles represented by 5.00×10^3 g of $(NH_4)_2SO_4$. Using atomic mass values from the periodic table, you can compute the molar mass of $(NH_4)_2SO_4$ to be 132.14 g/mol.

$$n_{(NH_4)_2SO_4} = \frac{5.00 \times 10^3 \; \cancel{g} \; (NH_4)_2SO_4}{132.14 \; \cancel{g}/mol \; (NH_4)_2SO_4} = 37.84 \; mol \; (NH_4)_2SO_4$$

Next, determine the number of moles of NH_3 that must react to produce 37.84 mol $(NH_4)_2SO_4$.

$$n_{NH_3} = 37.84 \; \cancel{mol \; (NH_4)_2SO_4} \times \frac{2 \; mol \; NH_3}{1 \; \cancel{mol \; (NH_4)_2SO_4}} = 75.68 \; mol \; NH_3$$

Finally, use the ideal gas law equation to calculate the volume of 75.68 mol NH_3 under the stated conditions. Solve the equation for V, the volume to be calculated.

$$PV = nRT; \quad V = \frac{nRT}{P}$$

Convert the temperature to kelvins, substitute known quantities into the equation, and compute the volume.

$$V = \frac{nRT}{P} = \frac{75.68 \; \cancel{mol} \; NH_3 \times 0.0821 \frac{L \cdot \cancel{atm}}{\cancel{mol} \cdot \cancel{K}} \times 351 \; \cancel{K}}{1.66 \; \cancel{atm}} = 1310 \; L$$

Notice that the values for the molar mass of $(NH_4)_2SO_4$ and the number of moles of NH_3 have more than three significant figures, whereas the calculated volume has only three. When you do a problem in a stepwise way, you should maintain at least one extra significant figure in the intermediate values you calculate. Then, round off values only at the end of the problem.

Practice Problems

24. A piece of aluminum with a mass of 4.25 g is used to generate hydrogen gas by the following method.

$$2Al(s) + 6HCl(aq) \rightarrow 2AlCl_3(aq) + 3H_2(g)$$

The hydrogen is collected at a temperature of 15°C and a pressure of 94.4 kPa. What volume of hydrogen is produced?

25. Ammonium nitrate decomposes when heated to produce dinitrogen monoxide and water.

$$NH_4NO_3(s) \rightarrow N_2O(g) + 2H_2O(l)$$

If 12.2 g of NH_4NO_3 reacts, what volume of NO, measured at 98.5 kPa and 14°C, is produced?

26. Carbon disulfide burns in air to produce carbon dioxide gas and sulfur dioxide gas.

$$CS_2(l) + 3O_2(g) \rightarrow CO_2(g) + 2SO_2(g)$$

Determine the mass of CS_2 required to produce 15.7 L of SO_2 gas at 25°C and 99.4 kPa. (Hint: This problem is the reverse of Example Problem 14-7. Start by calculating the moles of SO_2 present.)

27. When potassium chlorate is heated, it decomposes to produce potassium chloride and oxygen gas. Write a balanced equation for this reaction, and calculate the mass of potassium chlorate needed to produce 5.00×10^2 mL of oxygen gas at 1.108 atm and 39°C.

Chapter 14 Review

28. Describe the spacing of the particles of an ideal gas and the nature of the collisions of these particles.

29. Why does the pressure of a gas increase when the gas is squeezed into a smaller volume?

30. A sample of gas is squeezed until its pressure triples. How is the volume of the gas changed by the squeezing?

31. A sample of gas is heated until its Kelvin temperature doubles. How is the sample's volume changed by the heating?

32. How many moles of gas are there in 1000.0 L of nitrogen gas measured at STP?

33. In the reaction $N_2(g) + 3H_2(g) \rightarrow 2NH_3(g)$, what volume of ammonia is formed when 360 mL of H_2 gas reacts? Assume that both gases are at STP.

Solutions

15.1 What are solutions?

Solutions are homogeneous mixtures that contain a solute and a solvent. Solutions, solutes, and solvents may be solid, liquid, or gaseous. A substance that dissolves in a solvent is said to be **soluble** in that solvent. A substance that does not dissolve in a solvent is said to be **insoluble** in that solvent. Two liquids that are soluble in each other are said to be **miscible.** Liquids that are not soluble in each other are **immiscible.**

A soluble substance is able to dissolve in a solvent because attractive forces between solvent and solute particles are strong enough to overcome the attractive forces holding the solute particles together. Solvent particles surround solute particles to form a solution in a process called **solvation.** Solvation in water is also called hydration.

In general, "like dissolves like"; that is, polar substances tend to be soluble in polar solvents, and nonpolar substances tend to be soluble in nonpolar solvents. When an ionic compound dissolves in water, a polar solvent, the attractions between the water dipoles and the ions cause the ions to become solvated. The ions break away from the surface of the ionic solid and move into solution. The overall energy change that occurs during solution formation is called the **heat of solution.**

There are three common ways to increase the collisions between solute and solvent particles and thus increase the rate at which the solute dissolves: agitating the mixture, increasing the surface area of the solute, and increasing the temperature of the solvent. Increasing the temperature causes collisions to become more frequent and to have greater energy.

The **solubility** of a solute is the maximum amount of the solute that will dissolve in a given amount of solvent at a specified temperature and pressure. Solubility is usually expressed in grams of solute per 100 g of solvent. A solution that contains the maximum amount of dissolved solute for a given amount of solvent at a specific temperature and pressure is called a **saturated solution.** An

unsaturated solution is one that contains less dissolved solute for a given temperature and pressure than does a saturated solution. A **supersaturated** solution contains more dissolved solute than does a saturated solution at the same temperature. Supersaturated solutions are unstable, and the excess solute they contain often precipitates out of solution if the solution is disturbed.

Various factors affect solubility. For example, most solids are more soluble at higher temperatures than at lower ones, as shown in the graph below, which plots solubility versus Celsius temperature for a number of solutes.

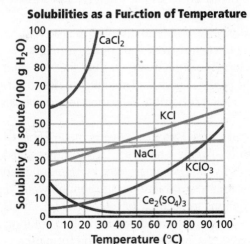

Solubilities as a Function of Temperature

Example Problem 15-1
Determining Solubility

Use the graph above to determine the approximate solubility of potassium chloride (KCl) at 20°C. How many grams of KCl would dissolve in 300 g of water at that same temperature? How many more grams of KCl could be dissolved if the temperature of the water was raised to 40°C?

Locate the curve representing KCl on the graph. At 20°C, the solubility of KCl is approximately 33 g KCl per 100 g water. Using this solubility, calculate the amount of KCl that will dissolve in 300 g of water at 20°C.

$$\frac{33 \text{ g KCl}}{100 \text{ g H}_2\text{O}} \times 300 \text{ g H}_2\text{O} = 99 \text{ g KCl}$$

According to the graph, the solubility of KCl at 40°C is 40 g KCl per 100 g H_2O. Find the amount of KCl that will dissolve in 300 g of water at 40°C.

$$\frac{40 \text{ g KCl}}{100 \text{ g H}_2\text{O}} \times 300 \text{ g H}_2\text{O} = 120 \text{ g KCl}$$

The amount of additional KCl that dissolves at the higher temperature in 300 g of water is the difference between the two amounts.

$$120 \text{ g KCl} - 99 \text{ g KCl} = 21 \text{ g KCl}$$

Practice Problems

1. Use the graph on the previous page to determine each of the following.
   a. the solubility of $CaCl_2$ at 5°C
   b. the solubility of $CaCl_2$ at 25°C
   c. the mass of $Ce_2(SO_4)_3$ that will dissolve in 50 g of water at 10°C
   d. the mass of $Ce_2(SO_4)_3$ that will dissolve in 50 g of water at 0°C
   e. the mass of additional $KClO_3$ that will dissolve in 200 g of water if the water temperature is raised from 30°C to 90°C

▶ **Pressure and solubility** The solubility of a gas increases as its external pressure (its pressure above the solution) increases. This fact is expressed by **Henry's law,** which states that at a given temperature, the solubility (S) of a gas in a liquid is directly proportional to the pressure (P) of the gas above the liquid, or

$$\frac{S_1}{P_1} = \frac{S_2}{P_2}$$

where S_1 is the solubility of a gas at pressure P_1, and S_2 is the solubility of the gas at P_2.

Example Problem 15-2
Using Henry's Law

If 0.24 g of a gas dissolves in 1.0 L of water at 1.5 atm of pressure, how much of the gas will dissolve if the pressure is raised to 6.0 atm? Assume the temperature is held constant.

In this case, $S_1 = 0.24$ g/L, $P_1 = 1.5$ atm, and $P_2 = 6.0$ atm. Use Henry's law to calculate S_2.

Rearrange Henry's law to solve for S_2, then substitute the known values into the equation and solve.

$$S_2 = S_1\left(\frac{P_2}{P_1}\right) = (0.24 \text{ g/L})\left(\frac{6.0 \text{ atm}}{1.5 \text{ atm}}\right) = 0.96 \text{ g/L}$$

Practice Problems

2. A gas has a solubility of 0.086 g/L at a pressure of 3.5 atm. At what pressure would its solubility be 2.3 g/L?

3. The solubility of a gas changes from 0.95 g/L to 0.72 g/L. If the initial pressure was 2.8 atm, what is the final pressure?

15.2 Solution Concentration

The **concentration** of a solution is a measure of the amount of solute dissolved in a given amount of solvent or solution. A concentrated solution contains a large amount of solute. A dilute solution contains a small amount of solute.

▶ **Using percent to describe concentration** It is useful to describe solution concentrations quantitatively, in terms of relative masses or volumes. One method is percent by mass, which is the ratio of the solute's mass to the solution's mass, expressed as a percent.

$$\text{Percent by mass} = \frac{\text{mass of solute}}{\text{mass of solution}} \times 100$$

Example Problem 15-3
Determining Percent by Mass

What is the percent by mass of potassium nitrate in a solution made by mixing 5.4 g of this substance with 260.0 mL of water?

The mass in grams of water is numerically equal to the volume of the water in milliliters, because 1 mL water has a mass of 1 g. Thus, the mass of the solvent (water) is 260.0 g. The mass of the solution is the sum of the masses of solute and solvent.

$$\text{Mass of solution} = 5.4 \text{ g} + 260.0 \text{ g} = 265.4 \text{ g}$$

Substitute the mass of solution and the given mass of solute into the equation for percent by mass and solve.

$$\text{Percent by mass} = \frac{\text{mass of solute}}{\text{mass of solution}} \times 100$$

$$= \frac{5.4 \text{ g}}{265.4 \text{ g}} \times 100 = 2.0\%$$

Practice Problems

4. What is the percent by mass of sodium carbonate in a water solution containing 0.497 g $NaCO_3$ in 58.3 g of solution?

5. The percent by mass of magnesium chloride in a water solution is 1.47%. How many grams of solute are dissolved in each 500.00 g of solution?

6. What is the mass of the solvent in Practice Problem 5?

Percent by volume is a method of expressing the concentration of solute in a solution made by mixing two liquids.

$$\text{Percent by volume} = \frac{\text{volume of solute}}{\text{volume of solution}} \times 100$$

For example, 100 mL of a solution made by mixing water with 50 mL of acetone has a percent by volume of acetone of (50 mL/100 mL) × 100, or 50%.

Practice Problems

7. What is the percent by volume of isopropyl alcohol in a solution made by mixing 75 mL of the alcohol with enough water to make 288 mL of solution?

8. What volume of acetone was used to make 3.11 L of a water solution if the percent acetone by volume is 27.9%?

▶ **Molarity** A common way to express solution concentrations is **molarity** (M), the number of moles of solute per liter of solution.

$$\text{Molarity } (M) = \frac{\text{moles of solute}}{\text{liter of solution}}$$

Example Problem 15-4
Calculating Molarity and Preparing Molar Solutions

a. What is the molarity of an aqueous solution that contains 14.2 g NaCl dissolved in 2364 mL of the solution?

b. How many grams of NaCl are contained in 1.00 L of this solution?

a. Use molar mass to convert grams of solute to moles of solute.

$$(14.2 \text{ g NaCl})\left(\frac{1 \text{ mol NaCl}}{58.5 \text{ g NaCl}}\right) = 0.243 \text{ mol NaCl}$$

The solution's volume must be expressed in liters.

$$(2364 \text{ mL})\left(\frac{1 \text{ L}}{1000 \text{ mL}}\right) = 2.364 \text{ L}$$

Substitute the known values and solve.

$$\text{Molarity } (M) = \frac{\text{moles of solute}}{\text{liter of solution}} = \frac{0.243 \text{ mol NaCl}}{2.364 \text{ L solution}}$$

$$= \frac{0.103 \text{ mol NaCl}}{\text{L solution}} = 0.103M$$

b. Given the definition of molarity, it is now known that there is 0.103 mole of NaCl per liter of this solution. To find the number of grams this represents, multiply by the molar mass of NaCl.

$$\frac{0.103 \text{ mol NaCl}}{1 \text{ L solution}} \times \frac{58.5 \text{ g NaCl}}{1 \text{ mol NaCl}} = \frac{6.03 \text{ g NaCl}}{1 \text{ L solution}}$$

Practice Problems

9. A solution is made by dissolving 17.0 g of lithium iodide (LiI) in enough water to make 387 mL of solution. What is the molarity of the solution?

10. Calculate the molarity of a water solution of $CaCl_2$, given that 5.04 L of the solution contains 612 g of $CaCl_2$.

▶ **Diluting molar solutions** Suppose you wished to dilute a stock solution of known concentration to make a given quantity of solution of lower concentration? You would need to know the volume of stock solution to use. To find that information, you can use the equation

$$M_1V_1 = M_2V_2$$

where M_1 and V_1 are the molarity and volume, respectively, of the stock solution, and M_2 and V_2 are the molarity and volume, respectively, of the dilute solution. The equation can be solved for any of the four values, given the other three.

Example Problem 15-5
Diluting a Solution

What volume, in milliliters, of a 1.15M stock solution of potassium nitrate is needed to make 0.75 L of 0.578M potassium nitrate?

Use the following equation.

$$M_1V_1 = M_2V_2$$

Rearrange the equation to solve for the volume of stock solution, V_1, and substitute the known values into the equation.

$$V_1 = V_2\left(\frac{M_2}{M_1}\right) = (0.75\text{L})\left(\frac{0.578M}{1.15M}\right) = 0.377 \text{ L}$$

Convert to milliliters.

$$(0.377 \text{ L})\left(\frac{1000 \text{ mL}}{\text{L}}\right) = 377 \text{ mL}$$

Thus, in making the 0.578M solution, 377 mL of the 1.15M stock solution should be diluted with enough water to make 0.75 L of solution.

Practice Problems

11. Suppose you wished to make 0.879 L of 0.250M silver nitrate by diluting a stock solution of 0.675M silver nitrate. How many milliliters of the stock solution would you need to use?

12. If 55.0 mL of a 2.45M stock solution of sucrose is diluted with water to make 168 mL of sucrose solution, what is the molarity of the final solution?

▶ **Molality** The **molality** (m) of a solution is equal to the number of moles of solute per kilogram of solvent.

$$\text{Molality } (m) = \frac{\text{moles of solute}}{\text{kilogram of solvent}}$$

Example Problem 15-6
Calculating Molality

What is the molality of a solution that contains 16.3 g of potassium chloride dissolved in 845 g of water?

Convert the mass of solute to moles.

$$(16.3 \text{ g KCl})\left(\frac{1 \text{ mol KCl}}{74.6 \text{ g KCl}}\right) = 0.218 \text{ mol KCl}$$

The solvent mass, 845 g, must be expressed in kilograms.

$$(845 \text{ g H}_2\text{O})\left(\frac{1 \text{ kg H}_2\text{O}}{1000 \text{ g H}_2\text{O}}\right) = 0.845 \text{ kg H}_2\text{O}$$

Substitute the known values into the equation for molality and solve.

$$\text{Molality } (m) = \frac{\text{moles of solute}}{\text{kilogram of solvent}}$$

$$= \frac{0.218 \text{ mol KCl}}{0.845 \text{ kg water}} = \frac{0.258 \text{ mol KCl}}{\text{kg water}} = 0.258m$$

Practice Problems

13. What is the molality of the solution formed by mixing 104 g of silver nitrate ($AgNO_3$) with 1.75 kg of water?

14. Suppose that 5.25 g of sulfur (S_8) is dissolved in 682 g of the liquid solvent carbon disulfide (CS_2). What is the molality of the sulfur solution?

▶ **Mole fraction** There is yet another way of expressing solution concentration. A **mole fraction** is the ratio of the number of moles of solute or solvent to the total number of moles of solute plus solvent in a solution. The mole fraction for the solvent (X_A) and the mole fraction for the solute (X_B) are equal to

$$X_A = \frac{n_A}{n_A + n_B} \qquad X_B = \frac{n_B}{n_A + n_B}$$

where n_A is the number of moles of solvent and n_B is the number of moles of solute.

For example, suppose you wish to find the mole fraction of the solvent and of the solute in a solution that contains 215 g of water and 44.0 g of sodium hydroxide (NaOH). First determine the number of moles of solute and solvent by dividing each mass by its molar mass.

$$n_{H_2O} = 215 \text{ g } H_2O \times \frac{1 \text{ mol } H_2O}{18.0 \text{ g } H_2O} = 11.9 \text{ mol } H_2O$$

$$n_{NaOH} = 44.0 \text{ g NaOH} \times \frac{1 \text{ mol NaOH}}{40.0 \text{ g NaOH}} = 1.10 \text{ mol NaOH}$$

Then substitute the values into the formula for mole fraction.

$$X_{H_2O} = \frac{n_{H_2O}}{n_{H_2O} + n_{NaOH}} = \frac{11.9 \text{ mol}}{11.9 \text{ mol} + 1.10 \text{ mol}} = 0.915$$

$$X_{NaOH} = \frac{n_{NaOH}}{nH_2O + n_{NaOH}} = \frac{1.10 \text{ mol}}{11.9 \text{ mol} + 1.10 \text{ mol}} = 0.0846$$

Practice Problems

15. Determine the mole fraction of the solvent and solute in each of the following solutions.

a. 855 g water, 448 g ethanol (C_2H_5OH)

b. 761.0 g water, 70.0 g calcium chloride ($CaCl_2$)

c. 945 g carbon tetrachloride (CCl_4), 265 g benzene (C_6H_6)

15.3 Colligative Properties of Solutions

Physical properties of a solution that are affected by the number of solute particles but not by the identity of those particles are called

colligative properties. One colligative property is **vapor pressure lowering.** The vapor pressure of a solution containing a nonvolatile solute is lower than the vapor pressure of the pure solvent. This is because in the solution there are fewer solvent molecules at the surface from which evaporation takes place. Thus, there are fewer solvent molecules escaping and less vapor produced, thereby lowering the vapor pressure. The greater the number of solute particles in the solvent, the lower the vapor pressure.

The effects of solutes on colligative properties depend upon the actual concentration of solute particles. For nonelectrolytes, the solute particle concentration is the same as the solute concentration. This is because nonelectrolytes, such as sucrose, do not ionize in solution. Electrolytes are substances that ionize or that dissociate into ions. Thus, electrolytes produce particle concentrations higher than those of the original substance. For example, sodium chloride, $NaCl$, dissociates almost completely into separate sodium ions and chloride ions, so a $1m$ $NaCl$ solution is actually nearly $2m$ in particles.

Two other colligative properties are the raising of a solution's boiling point and the reduction of its freezing point. The temperature difference between the boiling point of a solution and the boiling point of its pure solvent is called **boiling point elevation.** The greater the number of solute particles in the solution, the greater is the boiling point elevation. Boiling point elevation is symbolized ΔT_b and is related to the solution's molality as follows

$$\Delta T_b = K_b m$$

where m is the molality and K_b, the molal boiling point elevation constant, has a value that depends on the particular solvent. For water, K_b is equal to $0.512°C/m$.

The temperature difference between the freezing point of a solution and the freezing point of its pure solvent is called **freezing point depression.** The freezing point of a solution is always lower than that of a pure solvent. Freezing point depression is symbolized ΔT_f and is related to the solution's molality as follows

$$\Delta T_f = K_f m$$

where m is the molality and K_f, the molal freezing point depression constant, has a value that depends on the particular solvent. For water, K_f is equal to $1.86°C/m$.

Example Problem 15-7
Changes in Boiling and Freezing Points

What are the boiling point and freezing point of a 0.750*m* aqueous solution of the electrolyte potassium bromide (KBr)?

Find the effective solute particle concentration. Each formula unit of KBr dissociates into two ions, so the given molality must be doubled.

$$2 \times 0.750m = 1.50m$$

Substitute the known values and solve.

$$\Delta T_b = K_b m = (0.512°C/m) \times 1.50m = 0.768°C$$

The solution's boiling point is 100.0°C + 0.768°C = 100.768°C.

Substitute the known values and solve.

$$\Delta T_f = K_f m = (1.86°C/m) \times 1.50m = 2.79°C$$

The solution's freezing point is 0.0°C − 2.79°C = −2.79°C.

Practice Problems

16. What are the boiling and freezing points of a 1.34*m* water solution of calcium chloride ($CaCl_2$)?

17. Calculate the molality of a water solution of a nonelectrolyte, given that the freezing point depression of the solution is 4.33°C.

Another colligative property is **osmotic pressure,** a pressure related to a process called **osmosis.** Osmosis is the diffusion of solvent particles across a semipermeable membrane from an area of lower solute concentration to an area of higher solute concentration. The osmotic pressure effect is due to the fact that a smaller amount of solute on the less concentrated side of the membrane results in a larger number of solvent molecules available at that surface of the membrane. Thus, there is more diffusion of solvent from the side of lower solute concentration.

15.4 Heterogeneous Mixtures

In a true solution, which is a homogeneous mixture, the solute particles are less than 1 nm in diameter and remain distributed throughout the solution in a single phase. In contrast, a heterogeneous mixture contains materials that exist in distinct phases. In one

type of heterogeneous mixture, a **suspension,** the suspended particles are greater than 1000 nm in diameter and eventually settle out
of the mixture. In another type of heterogeneous mixture, a **colloid,**
the particles are between 1 nm and 1000 nm in diameter. Normally,
they do not settle out of the mixture, due in part to **Brownian
motion,** an erratic movement of particles. Colloids that appear clear
can be distinguished from solutions because the particles in the
colloids are large enough to scatter light, a phenomenon called
the **Tyndall effect.**

Practice Problems

18. Four mixtures having the following properties are examined.
Identify each mixture as a solution, a suspension, or a colloid.
If the properties are not conclusive for identification, state
which types of mixtures are possible.

   a. quickly settles out into a watery liquid and an oily liquid

   b. is transparent and does not settle out

   c. is transparent, scatters the light from a flashlight beam, and
shows zigzag motions when observed through a microscope

   d. is transparent and does not scatter the light from a flashlight
beam

Chapter 15 Review

19. Define *solvation*. Explain what happens in terms of attractive
forces when a substance dissolves in a solvent.

20. Contrast saturated, unsaturated, and supersaturated solutions.

21. State Henry's law in words, and write the equation for it.

22. Explain how percent by mass is calculated.

23. What is the difference between a $2M$ solution and a $2m$
solution?

24. Define *colligative property*, and identify four colligative
properties.

25. Would you expect the freezing point depression of a $1m$-solution
of sucrose and that of a $1m$-solution of sodium chloride to be
the same? Explain.

26. Contrast a solution, a suspension, and a colloid in terms of their
nature, properties, and particle size.

Energy and Chemical Change

16.1 Energy

Energy is the ability to do work or produce heat. Heat is commonly measured in joules or calories. One calorie is equivalent to 4.184 joules.

The amount of heat required to raise the temperature of one gram of a substance by one degree Celsius is the **specific heat** of the substance. Liquid water has a high specific heat of 4.184 J/(g·°C). By contrast, the specific heat of iron is 0.449 J/(g·°C). When the temperature of a substance changes, the amount of heat absorbed or released is given by the following equation.

$$q = c \times m \times \Delta T$$

In the equation, q = the heat absorbed or released, c = the specific heat of the substance, m = the mass of the sample in grams, and ΔT is the change in temperature in °C.

Example Problem 16-1
Calculating Heat

A silver bar with a mass of 250.0 g is heated from 22.0°C to 68.5°C. How much heat does the silver bar absorb?

Use the equation for heat.

$$q = c \times m \times \Delta T$$

The temperature change is the difference between the final temperature and the initial temperature.

$$\Delta T = 68.5°C - 22.0°C = 46.5°C$$

From Table 16-2 in your textbook, the specific heat of silver is 0.235 J/(g·°C). Substitute the known values to solve for the amount of heat absorbed.

$$q = 0.235 \text{ J/(g·°C)} \times 250.0 \text{ g} \times 46.5°C = 2730 \text{ J}$$

Practice Problems

1. How much heat does a 23.0-g ice cube absorb as its temperature increases from $-17.4°C$ to $0.0°C$? Give the answer in both joules and calories.

2. A sample of an unknown metal has a mass of 120.7 g. As the sample cools from $90.5°C$ to $25.7°C$, it releases 7020 J of energy. What is the specific heat of the sample? Identify the metal among those in Table 16-2 in your textbook.

3. A 15.6-g sample of ethanol absorbs 868 J as it is heated. If the initial temperature of the ethanol was $21.5°C$, what is the final temperature of the ethanol?

16.2 Heat in Chemical Reactions and Processes

Heat changes that occur during chemical and physical processes can be measured using an insulated device called a **calorimeter.** In one type of calorimeter, the temperature change of a known mass of water is used to determine the amount of energy released or absorbed by a system undergoing a chemical or physical change. The following example problem shows you how to determine the specific heat of an unknown substance by using calorimetry data.

Example Problem 16-2
Using Data from Calorimetry

A calorimeter contains 195 g of water at $20.4°C$. A 37.8-g sample of an unknown metal is heated to $133°C$ and placed into the water in the calorimeter. Heat flows from the metal to the water until both reach a final temperature of $24.6°C$. What is the specific heat of the metal?

First, calculate the heat gained by the water.

$$q_w = c_w \times m_w \times \Delta T_w$$
$$q_w = 4.184 \text{ J/(g·°C)} \times 195 \text{ g} \times (24.6°C - 20.4°C)$$
$$q_w = 4.184 \text{ J/(g·°C)} \times 195 \text{ g} \times 4.2°C$$
$$q_w = 3430 \text{ J}$$

Note that three significant figures are retained for precision in the calculation. At the end of the problem, you should round off the final answer to two significant figures.

In the calorimeter, the amount of heat gained by the water equals the amount of heat lost by the metal sample.

$$q = q_w = q_m = 3430 \text{ J}$$

Now examine the equation for the heat lost by the metal.

$$q = c_m \times m_m \times \Delta T_m$$

This equation can be rearranged to find the specific heat of the metal.

$$c_m = \frac{q}{m_m \times \Delta T_m}$$

The temperature change for the metal is as follows.

$$\Delta T_m = 133°C - 24.6°C = 108°C$$

Substitute the known values into the equation for the specific heat of the metal.

$$c_m = \frac{q}{m_m \times \Delta T_m} = \frac{3430 \text{ J}}{37.8 \text{ g} \times 108°C} = 0.84 \text{ J/(g} \cdot °C)$$

Practice Problems

4. A 50.6-g sample of iron metal is heated and put into 104 g of water at 19.7°C in a calorimeter. If the final temperature of the iron sample and the water is 24.3°C, what was the temperature of the iron sample when it was placed in the water?

5. A 77.5-g sample of an unknown solid is heated to 62.5°C and placed into a calorimeter containing 93 g of water at 23.3°C. If the final temperature of the solid sample and the water is 26.2°C, what is the specific heat of the solid?

6. If 40.0 g of water at 70.0°C is mixed with 40.0 g of ethanol at 10.0°C, what is the final temperature of the mixture? (Hint: The heat lost by the water equals the heat gained by the ethanol. Assume no heat loss to the surroundings.)

Thermochemistry is the study of heat changes that accompany chemical reactions and phase changes. In thermochemistry, the **system** is the reaction or process being studied, and everything outside the system is called the **surroundings.** The **universe** is defined as the system plus the surroundings.

The heat content of a system at constant pressure is called the **enthalpy** (H) of the system. The heat absorbed or released during a change in a system at constant pressure is the change in enthalpy (ΔH). The enthalpy change for a chemical reaction is called the **enthalpy of reaction** or **heat of reaction** and is defined by this equation.

$$\Delta H_{rxn} = H_{products} - H_{reactants}$$

In an endothermic reaction, heat is absorbed and ΔH_{rxn} is positive. In an exothermic reaction, heat is released and ΔH_{rxn} is negative.

16.3 Thermochemical Equations

A **thermochemical equation** is a balanced chemical equation that includes the physical states of the reactants and products and the change in enthalpy. For example, the thermochemical equation for the combustion of ethanol is as follows.

$$C_2H_5OH(l) + 3O_2(g) \rightarrow 2CO_2(g) + 3H_2O(l) \quad \Delta H_{comb} = -1367 \text{ kJ}$$

The enthalpy change for the complete burning of one mole of a substance is the **enthalpy (heat) of combustion** (ΔH_{comb}) for that substance. Heat is also absorbed or released during changes of state. The heat required to vaporize one mole of a liquid is called its **molar enthalpy (heat) of vaporization** (ΔH_{vap}). The heat required to melt one mole of a solid is its molar enthalpy (heat) of fusion (ΔH_{fus}).

Example Problem 16-3
Calculating Enthalpy of Reaction

The enthalpy of combustion for methanol (CH_3OH) is -726 kJ/mol. How much heat is released when 82.1 g of methanol is burned?

The enthalpy of combustion is negative, so the reaction is exothermic and heat is released. The molar mass of methanol is 32.05 g/mol. First, calculate the number of moles of methanol that is burned.

$$82.1 \text{ g CH}_3\text{OH} \times \frac{1 \text{ mol CH}_3\text{OH}}{32.05 \text{ g CH}_3\text{OH}} = 2.56 \text{ mol CH}_3\text{OH}$$

Now find the enthalpy of reaction for the combustion of 82.1 g (2.56 mol) of methanol.

$$2.56 \text{ mol CH}_3\text{OH} \times \frac{(-726 \text{ kJ})}{1 \text{ mol CH}_3\text{OH}} = -1860 \text{ kJ}$$

Practice Problems

7. Calculate the heat required for the following two processes, and compare the results.

    a. A 100.0-g sample of solid ethanol melts at its melting point. $\Delta H_{\text{fus}} = 4.94$ kJ/mol

    b. A 100.0-g sample of liquid ethanol vaporizes at its boiling point. $\Delta H_{\text{vap}} = 38.6$ kJ/mol

8. How much heat is evolved when 24.9 g of propanol (C_3H_7OH) is burned? $\Delta H_{\text{comb}} = -2010$ kJ/mol

9. What mass of benzene (C_6H_6) must be burned in order to liberate 1.00×10^4 kJ of heat? $\Delta H_{\text{comb}} = -3268$ kJ/mol

16.4 Calculating Enthalpy Change

A theoretical way to determine ΔH for a chemical reaction is provided by **Hess's law,** which states that if two or more thermochemical equations can be added to produce a final equation for a reaction, then the enthalpy change for the final reaction equals the sum of the enthalpy changes for the individual reactions. The following example problem shows how to use Hess's law to find ΔH for a reaction.

Example Problem 16-4

Applying Hess's Law

Use thermochemical equations **a** and **b** to determine ΔH for the oxidation of ethanol (C_2H_5OH) to form acetaldehyde (C_2H_4O) and water.

$$2C_2H_5OH(l) + O_2(g) \rightarrow 2C_2H_4O(g) + 2H_2O(l)$$

a. $2C_2H_4O(g) + 5O_2(g) \rightarrow 4CO_2(g) + 4H_2O(l) \quad \Delta H = -2385$ kJ

b. $C_2H_5OH(l) + 3O_2(g) \rightarrow 2CO_2(g) + 3H_2O(l) \quad \Delta H = -1367$ kJ

Acetaldehyde should be on the right side of the equation, so reverse equation **a**. Note that you must change the sign of ΔH.

$$4CO_2(g) + 4H_2O(l) \rightarrow 2C_2H_4O(g) + 5O_2(g) \quad \Delta H = 2385 \text{ kJ}$$

The desired equation has two moles of ethanol, so double equation **b** and its ΔH.

$$2C_2H_5OH(l) + 6O_2(g) \rightarrow 4CO_2(g) + 6H_2O(l) \quad \Delta H = -2734 \text{ kJ}$$

Add these two equations, and cancel any terms common to both sides of the combined equation.

$$\cancel{4CO_2(g)} + \cancel{4H_2O(l)} \rightarrow 2C_2H_4O(g) + \cancel{5O_2(g)} \qquad \Delta H = 2385 \text{ kJ}$$

$$2C_2H_5OH(l) + \cancel{6}O_2(g) \rightarrow \cancel{4CO_2(g)} + \cancel{6}^{2}H_2O(l) \qquad \Delta H = -2734 \text{ kJ}$$

$$\overline{2C_2H_5OH(l) + O_2(g) \rightarrow 2C_2H_4O(g) + 2H_2O(l) \qquad \Delta H = -349 \text{ kJ}}$$

Note that ΔH is negative, and the reaction is moderately exothermic.

Practice Problems

10. Use reactions **a** and **b** to determine ΔH for this single-displacement reaction.

$$Cl_2(g) + 2HBr(g) \rightarrow 2HCl(g) + Br_2(g)$$

a. $H_2(g) + Cl_2(g) \rightarrow 2HCl(g) \quad \Delta H = -185 \text{ kJ}$
b. $H_2(g) + Br_2(g) \rightarrow 2HBr(g) \quad \Delta H = -73 \text{ kJ}$

11. Use reactions **a, b,** and **c** to determine ΔH for the reaction of carbon monoxide and hydrogen to form methanol (CH_3OH).

$$CO(g) + 2H_2(g) \rightarrow CH_3OH(l)$$

a. $2CO(g) + O_2(g) \rightarrow 2CO_2(g) \quad \Delta H = -566 \text{ kJ}$
b. $2H_2(g) + O_2(g) \rightarrow 2H_2O(l) \quad \Delta H = -572 \text{ kJ}$
c. $2CH_3OH(l) + 3O_2(g) \rightarrow 2CO_2(g) + 4H_2O(l)$
$\quad \Delta H = -1452 \text{ kJ}$

(Hint: First find ΔH for the reaction $2CO(g) + 4H_2(g) \rightarrow 2CH_3OH(l)$, then divide this result by 2 to obtain your final answer.)

▶ **Standard enthalpy (heat) of formation** The standard state of a substance is the normal state of the substance at 298 K (25°C) and one atmosphere pressure. For example, the standard state of water is a liquid, whereas the standard state of helium is a monatomic gas. The change in enthalpy that accompanies the formation of one mole of a compound in its standard state from its constituent elements in their standard states is called the **standard enthalpy (heat) of formation** (ΔH_f°) of the compound. For example, the standard heat of formation for magnesium oxide is given by this equation.

$$Mg(s) + \frac{1}{2}O_2(g) \rightarrow MgO(s) \quad \Delta H_f^\circ = -602 \text{ kJ}$$

Standard enthalpies of formation may be used with Hess's law to calculate enthalpies of reaction under standard conditions (ΔH_{rxn}°). The following formula summarizes the procedure.

$$\Delta H_{rxn}^\circ = \Sigma \Delta H_f^\circ(\text{products}) - \Sigma \Delta H_f^\circ(\text{reactants})$$

The equation states that the standard heat of reaction equals the sum of the standard heats of formation of the products minus the sum of the standard heats of formation of the reactants. In this equation, the standard heat of formation of an element in its standard state is zero.

Example Problem 16-5
Enthalpy Change from Standard Enthalpies of Formation

Use standard enthalpies of formation from Table A-8 in the back of this book to calculate ΔH_{rxn}° for the reaction of calcium metal with water to form calcium hydroxide and hydrogen gas.

$$Ca(s) + 2H_2O(l) \rightarrow Ca(OH)_2(s) + H_2(g)$$

Use the following formula.

$$\Delta H_{rxn}^\circ = \Sigma \Delta H_f^\circ(\text{products}) - \Sigma \Delta H_f^\circ(\text{reactants})$$

Because gaseous hydrogen and solid calcium are both elements in their standard states, their heats of formation are zero. The heat of reaction then becomes as follows.

$$\Delta H_{rxn}^\circ = \Delta H_f^\circ(Ca(OH)_2) - (2)\Delta H_f^\circ(H_2O(l))$$

Substituting the values yields the following.

$$\Delta H^\circ_{rxn} = -986.09 \text{ kJ} - 2(-285.830 \text{ kJ})$$

$$\Delta H^\circ_{rxn} = -986.09 \text{ kJ} + 571.660 \text{ kJ} = -414.43 \text{ kJ}$$

The reaction of one mole of calcium metal with two moles of water releases 414 kJ.

Practice Problems

12. Use standard enthalpies of formation from Table A-8 in the back of this book to calculate ΔH°_{rxn} for each of these reactions.

a. $Hg_2Cl_2(s) \rightarrow HgCl_2(s) + Hg(l)$

b. $2HCl(aq) + Zn(s) \rightarrow ZnCl_2(aq) + H_2(g)$

c. $C_2H_4(g) + 3O_2(g) \rightarrow 2CO_2(g) + 2H_2O(g)$

d. $HCl(aq) + AgCN(s) \rightarrow HCN(aq) + AgCl(s)$

16.5 Reaction Spontaneity

Entropy (S) is a measure of the disorder or randomness of the particles that make up a system. Spontaneous processes always result in an increase in the entropy of the universe. The change in the entropy of a system is given by the following equation.

$$\Delta S_{system} = S_{products} - S_{reactants}$$

Whether ΔS_{system} is positive or negative can be predicted in some cases by examining the reaction or process. Several factors affect the change in entropy of a system.

- **Changes of state** Entropy increases when a solid changes to a liquid and when a liquid changes to a gas because these changes of state result in freer movement of the particles.

- **Dissolving of a gas in a solvent** When a gas is dissolved in a liquid or solid solvent, the motion and randomness of the particles are limited and the entropy of the gas decreases.

- **Change in the number of gaseous particles** When the number of gaseous particles increases, the entropy of the system usually increases because more random arrangements are possible.

- **Dissolving of a solid or liquid to form a solution** When solute particles become dispersed in a solvent, the disorder of the particles and the entropy of the system usually increase.

- **Change in temperature** A temperature increase results in increased disorder of the particles and an increase in entropy.

Practice Problems

13. Predict the sign of ΔS_{system} for each of these changes.

a. $O_2(g) \rightarrow O_2(aq)$

b. $C_6H_6(s) \rightarrow C_6H_6(l)$

c. $C(s) + CO_2(g) \rightarrow 2CO(g)$

d. $C_{12}H_{22}O_{11}(s) \rightarrow C_{12}H_{22}O_{11}(aq)$

e. $2NO_2(g) \rightarrow N_2O_4(g)$

▶ **Free energy** For a reaction or process occurring at constant temperature and pressure, the energy that is available to do work is the **free energy** (G). The change in free energy is related to the change in enthalpy and the change in entropy by the following equation.

$$\Delta G_{system} = \Delta H_{system} - T\Delta S_{system}$$

In this equation, T is the Kelvin temperature. If ΔG_{system} is negative, the reaction or process is spontaneous; if ΔG_{system} is positive, the reaction or process is nonspontaneous.

Example Problem 16-6
Determining Reaction Spontaneity

For a chemical reaction, $\Delta H_{system} = -81$ kJ and $\Delta S_{system} = -215$ J/K. Is the reaction spontaneous at 50°C?

Convert the temperature to kelvins.

$$T = (50 + 273) \text{ K} = 323 \text{ K}$$

Find ΔG_{system}.

$$\Delta G_{system} = \Delta H_{system} - T\Delta S_{system}$$

$$\Delta G_{system} = -81\ 000 \text{ J} - (323 \text{ K})(-215 \text{ J/K})$$

$$\Delta G_{system} = -81\ 000\ J - (-69\ 400\ J)$$

$$\Delta G_{system} = -12\ 000\ J,\ or\ -12\ kJ$$

Because ΔG_{system} is negative, the reaction is spontaneous.

Practice Problems

14. Calculate ΔG_{system} for each process, and state if the process is spontaneous or nonspontaneous.

  a. $\Delta H_{system} = 147$ kJ, $T = 422$ K, $\Delta S_{system} = -67$ J/K

  b. $\Delta H_{system} = -43$ kJ, $T = 21°C$, $\Delta S_{system} = -118$ J/K

  c. $\Delta H_{system} = 227$ kJ, $T = 574$ K, $\Delta S_{system} = 349$ J/K

15. If $\Delta G_{system} = 0$, a reaction is in a state of chemical equilibrium, in which the forward reaction and the reverse reaction occur at equal rates. At what temperature is a reaction at chemical equilibrium if $\Delta H_{system} = -75$ kJ and $\Delta S_{system} = -192$ J/K?

Chapter 16 Review

16. A sample of ethanol and a sample of aluminum have the same mass. Which sample absorbs more energy as it is heated from 25°C to 56°C? Explain your answer.

17. What do the terms *system*, *surroundings*, and *universe* mean in thermochemistry?

18. Describe two changes of state for which ΔH is positive and two changes of state for which ΔH is negative.

19. Explain what is meant by the standard heat of formation of ethane gas, (C_2H_6). Include in your explanation the equation for the formation of ethane.

20. For each of the following processes, the signs of ΔH_{system} and ΔS_{system} are known. Explain whether you have sufficient information to determine whether each process is spontaneous.

  a. $\Delta H_{system} > 0$, $\Delta S_{system} < 0$

  b. $\Delta H_{system} < 0$, $\Delta S_{system} < 0$

  c. $\Delta H_{system} < 0$, $\Delta S_{system} > 0$

Reaction Rates

17.1 A Model for Reaction Rates

To determine the average rate of a process, you must know how much a quantity changes over time. The Greek letter delta (Δ) is the mathematical symbol for "change in."

$$\text{Average rate of a process} = \frac{\text{change in quantity}}{\text{change in time}} = \frac{\Delta \text{ quantity}}{\Delta \text{ time}}$$

▶ **Average Reaction Rates** Consider a reaction between carbon monoxide and nitrogen dioxide gas to form nitrogen monoxide and carbon dioxide.

$$CO(g) + NO_2(g) \rightarrow NO(g) + CO_2(g)$$

The rate of this reaction can be expressed as the rate of disappearance of either of the reactants or the rate of appearance of either of the products. Suppose that at the beginning of the reaction, the reactant CO has a concentration of 0.0223 mol/L. After 12.5 s, the concentration of CO has dropped to 0.0119 mol/L. Because the amount of reactant decreases, the change in concentration will have a negative value, but the rate of a chemical reaction must have a positive value. Therefore, when a rate is determined by measuring the disappearance of a reactant, a minus sign is used in the expression.

$$\text{Average reaction rate} = -\frac{\Delta[CO]}{\Delta t} = -\left(\frac{[CO]_2 - [CO]_1}{t_2 - t_1} \right)$$

$$\text{Average reaction rate} = -\left(\frac{0.0119 \text{ mol/L} - 0.0223 \text{ mol/L}}{12.5 \text{ s} - 0 \text{ s}} \right)$$

$$\text{Average reaction rate} = \frac{-(-0.0104 \text{ mol/L})}{12.5 \text{ s}}$$

$$\text{Average reaction rate} = 0.000\,832 \frac{\text{mol}}{\text{L} \cdot \text{s}}$$

Example Problem 17-1
Calculating Average Reaction Rate

Nitrogen monoxide reacts with chlorine gas to form the gaseous substance nitrosyl chloride according to the following equation.

$$2NO(g) + Cl_2(g) \rightarrow 2NOCl(g)$$

At the beginning of the reaction, the concentration of chlorine gas is 0.006 40 mol/L. After 30.0 s, the chlorine concentration has decreased to 0.002 95 mol/L. Calculate the average rate of the reaction over this time in terms of the disappearance of chlorine. First, write the mathematical expression for the average rate based on the rate of disappearance of the reactant Cl_2.

$$\text{Average reaction rate} = -\frac{\Delta[Cl_2]}{\Delta t} = -\left(\frac{[Cl_2]_2 - [Cl_2]_1}{t_2 - t_1} \right)$$

Next, substitute the data into the equation.

$$\text{Average reaction rate} = -\left(\frac{[Cl_2]_2 - [Cl_2]_1}{t_2 - t_1} \right)$$

$$= -\left(\frac{0.002\ 95\ \text{mol/L} - 0.006\ 40\ \text{mol/L}}{30.0\ \text{s} - 0\ \text{s}} \right)$$

$$\text{Average reaction rate} = -\left(\frac{-0.003\ 45\ \text{mol/L}}{30.0\ \text{s}} \right)$$

$$= 0.000\ 115\ \frac{\text{mol}}{\text{L} \cdot \text{s}}\ Cl_2$$

Practice Problems

Use the following information to solve the problems below. In aqueous solution, bromine reacts with formic acid (HCOOH) to produce hydrogen bromide and carbon dioxide gas.

$$Br_2(aq) + HCOOH(aq) \rightarrow 2HBr(aq) + CO_2(g)$$

Experimental Data for $Br_2(aq) + HCOOH(aq) \rightarrow 2HBr(aq) + CO_2(g)$			
Time (s)	[Br$_2$] (mol/L)	[HCOOH] (mol/L)	[HBr] (mol/L)
0.0	0.0120	0.0120	0.0
150.0	0.007 10	0.007 10	0.009 80

1. Calculate the average reaction rate over 150 s expressed in terms of the disappearance of Br_2.
2. Calculate the average reaction rate over 150 s expressed in terms of the appearance of HBr.
3. Explain the difference between the rate calculated in problem 2 and the rate calculated in problem 1.

▶ **Collision Theory** What must happen in order for substances to react? According to the **collision theory,** atoms, ions, and molecules must collide with each other in order to react. The following three statements summarize the collision theory.

1. Particles must collide in order to react.
2. The particles must collide with the correct orientation.
3. The particles must collide with enough energy to form an unstable **activated complex,** also called a **transition state,** which is an intermediate particle made up of the joined reactants.

The minimum amount of energy that colliding particles must have in order to form an activated complex is called the **activation energy** of the reaction. Particles that collide with less than the activation energy cannot form an activated complex.

17.2 Factors Affecting Reaction Rates

Other than the requirement that substances must be able to react, what conditions affect the rates of chemical reactions? Recall the collision theory. In order to react, particles of reactants must have an opportunity to collide, they must collide in the right orientation, and they must collide with enough energy to form an activated complex.

▶ **Surface Area** One way to increase the opportunity for collisions is to increase the surface area of a solid reactant. For example, a solid chunk of iron will rust slowly, but the same amount of iron made into extra-fine iron wool will rust so rapidly in moist air that it gives off heat.

▶ **Concentration** Increasing the concentration of one or more reactants puts more particles of the reactants in the same space, thereby increasing the frequency of collisions. The extra-fine iron wool mentioned earlier burns more vigorously in 100 percent oxygen gas than in air, which is approximately 20 percent oxygen gas.

▶ **Temperature** Increasing the temperature of a reaction increases the frequency of collisions between particles, allowing more opportunities for reaction. Also, increasing temperature raises the average kinetic energy of particles. Thus, more of the collisions have enough energy to form an activated complex.

▶ **Catalysts** A **catalyst** is a substance that increases the rate of a chemical reaction without being consumed in the reaction. A **heterogeneous catalyst** is in a different physical state from the reactants and works by providing a surface that attracts particles of both reactants and products. This action reduces the activation energy needed for the reaction because the particles can interact with each other on the surface.

A **homogeneous catalyst** is in the same physical state, usually in solution along with the reactants. For example, strong acids, such as sulfuric acid, act as homogeneous catalysts by providing a supply of hydronium ions that can attach to certain kinds of molecules, breaking covalent bonds and then returning to solution.

Substances called **inhibitors** are compounds that slow down, or inhibit, chemical reactions. Compounds used as preservatives in foods, medicines, and other perishable products are examples of inhibitors.

Practice Problems
Use the factors discussed in this section to account for the following observations.

4. When two gases react, compressing the gases generally increases the rate of reaction.

5. The rate of gaseous reactions can also be increased by pumping more gas into the reaction container.

6. A pile of kindling consists of small sticks, wood shavings, and scraps of paper and burns rapidly, giving off enough heat to begin a fire in a pile of large logs. The logs burn much slower than the kindling.

7. A laboratory chemist places a flask in an ice bath to prevent the reaction in the flask from foaming over.

8. Molecules of unburned fuel in a car exhaust are oxidized as they pass over powdered palladium.

17.3 Reaction Rate Laws

As reactants are used up, their concentrations decrease, the number of collisions decreases, and the reaction rate slows. A **rate law** is a mathematical expression that relates reaction rate to concentrations of reactants. The **reaction order** mathematically defines the extent to which reaction rate depends on the concentrations of reactants. Because most chemical reactions involve two or more reactants, the rate law often includes the concentrations of all reactants.

▶ **Determining Rate Laws** Consider the following reaction and its experimentally determined rate law.

$$2NO_2(g) + F_2(g) \rightarrow 2NO_2F(g)$$

$$\text{Rate} = k[NO_2][F_2]$$

This rate equation shows that the rate depends on the concentrations of both NO_2 and F_2, each to the first power. In other words, the reaction is first order in NO_2 and first order in F_2. So, if $[NO_2]$ is doubled while $[F_2]$ remains the same, the rate doubles. Also, if $[F_2]$ is doubled while $[NO_2]$ remains the same, the reaction rate doubles.

Now, examine the following reaction and its rate law.

$$2NO(g) + O_2(g) \rightarrow 2NO_2(g)$$

$$\text{Rate} = k[NO]^2[O_2]$$

Because the rate depends on the square of the concentration of NO, doubling [NO] while leaving $[O_2]$ the same will increase the reaction rate by a factor of 2^2, or 4.

Example Problem 17-2

Determining Rate Laws

Use the data in the table below to determine the form of the rate law for the following reaction.

$$2NO(g) + H_2(g) \rightarrow N_2(g) + 2H_2O(g)$$

Experimental Initial Rates for $2NO(g) + H_2(g) \rightarrow N_2(g) + 2H_2O(g)$		
Trial	Initial [NO] (mol/L)	Initial [H_2] (mol/L)	Initial rate of NO depletion $\frac{mol}{L \cdot s}$
1	4.0×10^{-3}	2.0×10^{-3}	1.2×10^{-5}
2	8.0×10^{-3}	2.0×10^{-3}	4.8×10^{-5}
3	4.0×10^{-3}	4.0×10^{-3}	2.4×10^{-5}

The general rate law for this type of reaction is as follows.

$$\text{Rate} = k[A]^m[B]^n$$

To start, compare the data from trials 1 and 2. Notice that [NO] in trial 1 is 4.0×10^{-3} mol/L and [NO] in trial 2 is 8.0×10^{-3}, double that in trial 1. Now see how the rate changes from trial 1 to trial 2. The rate in trial 2, 4.8×10^{-5} mol/L·s, is four times the rate in trial 1, which is 1.2×10^{-5} mol/L·s. When [NO] doubles, the initial reaction rate quadruples. Therefore, it is likely that the reaction rate depends on the square of the concentration of NO.

Next, determine how the rate depends on the change in [H_2]. When [H_2] doubles, the initial rate doubles. This result indicates that the rate is directly proportional to the concentration of H_2. Now you can write the rate law based on your comparisons.

$$\text{Rate} = k[NO]^2[H_2]^1 = k[NO]^2[H_2]$$

The rate law means that the reaction is second order in [NO] and first order in [H_2].

Practice Problems

9. The rate law for a reaction A + B → AB is rate = k[A][B]. What happens to the rate if the concentration of B is tripled while [A] remains the same? If the concentration of A is halved while [B] remains the same?

10. The rate law for a reaction 2X + Y → X_2Y is rate = k[X][Y]^2. What happens to the rate if the concentration of X is doubled while [Y] remains the same? If the concentration of Y is tripled while [X] remains the same?

11. Use the data in the following table to determine the reaction order with respect to [D] and [E]. Write the rate equation for the reaction.

Experimental Initial Rates for D + E → DE		
Trial	Initial [D] (mol/L)	Initial [E] (mol/L)	Initial rate of D depletion $\frac{mol}{L \cdot s}$
1	3.22×10^{-2}	8.68×10^{-3}	7.03×10^{-4}
2	6.44×10^{-2}	8.68×10^{-3}	2.81×10^{-3}
3	3.22×10^{-2}	1.74×10^{-2}	2.81×10^{-3}

17.4 Instantaneous Reaction Rates and Reaction Mechanisms

Many chemical reactions are complex. In a **complex reaction**, a certain series of steps must occur with correct fit and in an exact sequence in order to yield the reaction products. This series of steps is called a **reaction mechanism.**

As an example, consider that hydrogen peroxide (H_2O_2) decomposes when iodide ions (I^-) are present. Experiments have shown that the reaction takes place in two steps as follows.

Step 1	$H_2O_2 + I^- \rightarrow H_2O + IO^-$
Step 2	$H_2O_2 + IO^- \rightarrow H_2O + O_2 + I^-$
Net equation	$2H_2O_2 \rightarrow 2H_2O + O_2$

The IO^- ion is called an intermediate in the reaction. An **intermediate** is an atom, an ion, or a molecule produced in one step of a reaction and consumed in a later step. In a complex reaction, one step is always slower than the others. This step is called the **rate-determining step** because it will determine how fast the reaction forms products, no matter how fast the other steps are. In the above reaction, the first step is the slower step, therefore, it is the rate-determining step for the reaction $2H_2O_2 \rightarrow 2H_2O + O_2$.

Chapter 17 Review

12. Discuss two possible ways to determine the average rate of the following reaction in the laboratory. What quantities would you measure in each case?

$$H_2(g) + I_2(g) \rightarrow 2HI(g)$$

13. In the reaction $A + B \rightarrow C$, the concentration of A decreases from 6.27×10^{-2} mol/L to 4.75×10^{-2} mol/L in a span of 125 s. What is the rate of the reaction in terms of reactant A?

14. The collision theory says that particles must collide in order to react. What other two requirements must be met in order for particles to react?

15. Explain why increasing the concentration of a reactant usually increases the rate of a chemical reaction.

16. The rate law of the reaction $Q + R \rightarrow QR$ is rate $= k[Q]^2[R]$. How will the rate of the reaction change if $[Q]$ is doubled? Explain.

17. Explain how the rate-determining step of a complex reaction controls the overall reaction rate.

Chemical Equilibrium

18.1 Equilibrium: A State of Dynamic Balance

A **reversible reaction** is a reaction that can take place in both the forward and reverse directions. An example is the reaction of hydrogen and bromine at elevated temperature to form hydrogen bromide, and the reverse reaction, the decomposition of hydrogen bromide into its elements.

$$H_2(g) + Br_2(g) \rightarrow 2HBr(g)$$

$$2HBr(g) \rightarrow H_2(g) + Br_2(g)$$

These two equations may be combined into a single equation with a double arrow to show that both reactions occur.

$$H_2(g) + Br_2(g) \rightleftharpoons 2HBr(g)$$

When forward and reverse reactions occur at equal rates, a state of **chemical equilibrium** results in which the concentrations of reactants and products remain constant.

The **law of chemical equilibrium** states that at a given temperature, a chemical system may achieve a state in which a certain ratio of reactant and product concentrations has a constant value. The general equation for a reaction at equilibrium is as follows.

$$aA + bB \rightleftharpoons cC + dD$$

The law of chemical equilibrium may be used to write the equilibrium constant expression for the reaction.

$$K_{eq} = \frac{[C]^c[D]^d}{[A]^a[B]^b}$$

The **equilibrium constant,** K_{eq}, is the value of the ratio of the molar concentrations of the products divided by the molar concentrations of the reactants, with each concentration raised to the power equal to its coefficient in the balanced equation. For example, consider the reversible formation of hydrogen bromide.

$$H_2(g) + Br_2(g) \rightleftharpoons 2HBr(g)$$

This reaction is a **homogeneous equilibrium** because all the reactants and products are in the same physical state—they all are gases. Using the law of chemical equilibrium results in the following equilibrium constant expression.

$$K_{eq} = \frac{[HBr]^2}{[H_2][Br_2]}$$

The following example problem further demonstrates how to determine an equilibrium constant expression.

Example Problem 18-1
Equilibrium Constant Expressions for Homogeneous Equilibria

Write the equilibrium constant expression for the reaction of hydrogen sulfide and water vapor to form sulfur dioxide and hydrogen.

$$H_2S(g) + 2H_2O(g) \rightleftharpoons SO_2(g) + 3H_2(g)$$

All reactants and products are gases, so the equilibrium is homogeneous. Write a ratio in which the molar concentrations of the products are in the numerator and the molar concentrations of the reactants are in the denominator. Raise each concentration to its corresponding coefficient in the balanced chemical equation. The result is as follows.

$$K_{eq} = \frac{[SO_2][H_2]^3}{[H_2S][H_2O]^2}$$

Practice Problems

1. Write equilibrium constant expressions for the following homogeneous equilibria.

   a. $C_2H_4O(g) \rightleftharpoons CH_4(g) + CO(g)$

   b. $3O_2(g) \rightleftharpoons 2O_3(g)$

   c. $2N_2O(g) + O_2(g) \rightleftharpoons 4NO(g)$

   d. $4NH_3(g) + 3O_2(g) \rightleftharpoons 2N_2(g) + 6H_2O(g)$

An equilibrium in which the reactants and products of a reaction exist in more than one physical state is called a **heterogeneous equilibrium.** The equilibrium constant expression for a heterogeneous equilibrium is similar to that for a homogeneous equilibrium, except that the concentrations of pure solids and pure liquids are eliminated from the expression because the concentrations of these substances are constant values.

Example Problem 18-2
Equilibrium Constant Expressions for Heterogeneous Equilibria

Write the equilibrium constant expression for the high-temperature reaction of carbon dioxide and solid carbon to form carbon monoxide.

$$CO_2(g) + C(s) \rightleftharpoons 2CO(g)$$

Write a ratio with the concentration of the product in the numerator and the concentrations of the reactants in the denominator. Raise each concentration to the power equal to its coefficient in the balanced equation.

$$\frac{[CO]^2}{[CO_2][C]}$$

Leave out [C] because it is a pure solid with an unchanging concentration. The result is the equilibrium constant expression.

$$K_{eq} = \frac{[CO]^2}{[CO_2]}$$

Practice Problems

2. Write equilibrium constant expressions for the following heterogeneous equilibria.

   a. $C_4H_{10}(l) \rightleftharpoons C_4H_{10}(g)$
   b. $NH_4HS(s) \rightleftharpoons NH_3(g) + H_2S(g)$
   c. $CO(g) + Fe_3O_4(s) \rightleftharpoons CO_2(g) + 3FeO(s)$
   d. $(NH_4)_2CO_3(s) \rightleftharpoons 2NH_3(g) + CO_2(g) + H_2O(g)$

▶ **Calculating equilibrium constants** The value of K_{eq} is a constant for a given reaction at a given temperature. A K_{eq} value greater than 1 indicates that products are favored at equilibrium. If K_{eq} is less than 1, reactants are favored. The equilibrium concentrations of the reactants and products may be used to calculate K_{eq}, as shown in this example problem.

Example Problem 18-3

Calculating the Value of Equilibrium Constants

Nitrogen monoxide and bromine react to form nitrosyl bromide according to the following equation.

$$2NO(g) + Br_2(g) \rightleftharpoons 2NOBr(g)$$

Calculate K_{eq} for this equilibrium using the data [NOBr] = 0.0474 mol/L, [NO] = 0.312 mol/L, and [Br$_2$] = 0.259 mol/L.

The equilibrium constant expression is as follows.

$$K_{eq} = \frac{[NOBr]^2}{[NO]^2[Br_2]}$$

Substitute the known values into the equation and solve for K_{eq}.

$$K_{eq} = \frac{[0.0474]^2}{[0.312]^2[0.259]} = 0.0891$$

Note that K_{eq} is less than 1, so the reactants are favored in this equilibrium.

Practice Problems

3. The following is the chemical equation for the decomposition of formamide.

$$HCONH_2(g) \rightleftharpoons NH_3(g) + CO(g)$$

Calculate K_{eq} using the equilibrium data [HCONH$_2$] = 0.0637 mol/L, [NH$_3$] = 0.518 mol/L, and [CO] = 0.518 mol/L.

4. Hydrogen and carbon disulfide react to form methane and hydrogen sulfide according to this equation.

$$4H_2(g) + CS_2(g) \rightleftharpoons CH_4(g) + 2H_2S(g)$$

Calculate K_{eq} if the equilibrium concentrations are [H$_2$] = 0.205 mol/L, [CS$_2$] = 0.0664 mol/L, [CH$_4$] = 0.0196 mol/L, and [H$_2$S] = 0.0392 mol/L.

5. The equilibrium constant is 9.36 for the following reaction.

$$A(g) + 3B(g) \rightleftharpoons 2C(g)$$

The table below provides concentration data for two different reaction mixtures of these gases. Can you conclude that both reactions are at equilibrium? Explain your answer.

	Concentrations (mol/L)		
	A	**B**	**C**
Mixture 1	0.716	0.208	0.425
Mixture 2	0.562	0.491	0.789

18.2 Factors Affecting Chemical Equilibrium

Le Châtelier's principle states that if a stress is applied to a system at equilibrium, the system shifts in the direction that relieves the stress. For example, consider the equilibrium system in Example Problem 18-3.

$$2NO(g) + Br_2(g) \rightleftharpoons 2NOBr(g)$$

If an additional amount of reactant (NO or Br_2) is added to the system, the equilibrium will shift to the right, that is, more product (NOBr) will be formed. Conversely, adding more NOBr to the system will result in a shift to the left, forming more NO and Br_2. The removal of a reactant or product also results in a shift in the equilibrium. Removing a reactant causes the equilibrium to shift to the left, forming more reactants. Removing the product causes a shift to the right, forming more product.

▶ **Changes in volume** Le Châtelier's principle also applies to changes in the volume of the reaction vessel containing the equilibrium system. Suppose the volume of the reaction vessel for the $2NO(g) + Br_2(g) \rightleftharpoons 2NOBr(g)$ system is decreased, resulting in an increase in pressure. The equilibrium will shift to relieve the stress of increased pressure. In this case, the shift will be to the right because three moles of reactant gas combine to form only two moles of product gas. Thus, a shift toward the product will reduce the pressure of the system. If the volume of the reaction vessel was increased, the equilibrium would shift to the left, and more of the reactants would be formed.

Note that changing the volume of the reaction vessel causes no shift in the equilibrium when the number of moles of product gas equals the number of moles of reactant gas; an example is the equilibrium $H_2(g) + Br_2(g) \rightleftharpoons 2HBr(g)$ discussed in Section 18.1.

Practice Problems

6. Use Le Châtelier's principle to predict how each of the following changes would affect this equilibrium.

$$C_2H_4O(g) \rightleftharpoons CH_4(g) + CO(g)$$

a. adding $CH_4(g)$ to the system

b. removing $CO(g)$ from the system

c. removing $C_2H_4O(g)$ from the system

7. How would decreasing the volume of the reaction vessel affect these equilibria?

a. $CO(g) + H_2(g) \rightleftharpoons H_2CO(g)$

b. $NH_4HS(s) \rightleftharpoons NH_3(g) + H_2S(g)$

c. $2NbCl_4(g) \rightleftharpoons NbCl_3(g) + NbCl_5(g)$

d. $2SO_3(g) + CO_2(g) \rightleftharpoons CS_2(g) + 4O_2(g)$

▶ **Changes in temperature** Even though an equilibrium may shift to the right or left in response to a change in concentration or volume, the value of the equilibrium constant remains the same. A change in temperature, however, alters both the equilibrium position and the value of K_{eq}. For example, consider the thermochemical equation for the reversible formation of hydrogen chloride gas from its elements.

$$H_2(g) + Cl_2(g) \rightleftharpoons 2HCl(g) \quad \Delta H° = -185 \text{ kJ}$$

The forward reaction releases heat, so you can consider heat as a product in the forward reaction and a reactant in the reverse reaction.

$$H_2(g) + Cl_2(g) \rightleftharpoons 2HCl(g) + \text{heat}$$

Raising the temperature of this system requires the addition of heat, which shifts the equilibrium to the left and reduces the concentration of hydrogen chloride. Thus, the value of K_{eq} decreases. Lowering the temperature of the system means that heat is removed, so the equilibrium relieves the stress by shifting to the right, increasing both the concentration of hydrogen chloride and K_{eq}.

Practice Problems

8. Phosphorus pentachloride decomposes exothermically to form phosphorus trichloride and chlorine.

$$PCl_5(g) \rightleftharpoons PCl_3(g) + Cl_2(g) + heat$$

How would you regulate the temperature of this equilibrium in order to do the following?
a. increase the concentration of PCl_5
b. decrease the concentration of PCl_3
c. increase the amount of Cl_2 in the system
d. decrease K_{eq}

9. Predict how this equilibrium would respond to a simultaneous decrease in both temperature and pressure.

$$N_2O_4(g) \rightleftharpoons 2NO_2(g) \quad \Delta H° = +58 \text{ kJ}$$

18.3 Using Equilibrium Constants

When K_{eq} is known, the equilibrium concentration of a substance can be calculated if you know the concentrations of all other reactants and products. The following example problem shows you how to determine an equilibrium concentration.

Example Problem 18-4
Calculating Equilibrium Concentrations

At 350°C, $K_{eq} = 66.9$ for the formation of hydrogen iodide from its elements.

$$H_2(g) + I_2(g) \rightleftharpoons 2HI(g)$$

What is the concentration of HI if $[H_2] = 0.0295$ mol/L and $[I_2] = 0.0174$ mol/L?
Write the equilibrium constant expression.

$$\frac{[HI]^2}{[H_2][I_2]} = K_{eq}$$

Multiply both sides of the equation by $[H_2][I_2]$.

$$[HI]^2 = K_{eq} \times [H_2][I_2]$$

Substitute the known quantities into the equation and solve for $[HI]$.

$$[HI]^2 = 66.9 \times (0.0295)(0.0174) = 0.03434$$

An extra digit is retained here for accuracy, but the final answer will be rounded to three digits.

$$[HI] = \sqrt{0.03434} = 0.185 \text{ mol/L}$$

The equilibrium concentration of HI is 0.185 mol/L.

Practice Problems

10. At a certain temperature, $K_{eq} = 0.118$ for the following reaction.

$$2CH_4(g) \rightleftharpoons C_2H_2(g) + 3H_2(g)$$

Calculate these concentrations.

a. $[CH_4]$ in an equilibrium mixture with $[C_2H_2]$ = 0.0812 mol/L and $[H_2] = 0.373$ mol/L

b. $[C_2H_2]$ in an equilibrium mixture with $[CH_4] = 0.726$ mol/L and $[H_2] = 0.504$ mol/L

c. $[H_2]$ in an equilibrium mixture with $[CH_4] = 0.0492$ mol/L and $[C_2H_2] = 0.0755$ mol/L

11. A chemist studying the equilibrium $N_2O_4(g) \rightleftharpoons 2NO_2(g)$ controls the temperature so that $K_{eq} = 0.028$. At one equilibrium position, the concentration of N_2O_4 is 1.5 times greater than the concentration of NO_2. Find the concentrations of the two gases in mol/L. (Hint: Let $x = [NO_2]$ and $1.5x = [N_2O_4]$ in the equilibrium constant expression.)

▶ **Solubility equilibria** The **solubility product constant** (K_{sp}) is an equilibrium constant for the dissolving of a sparingly soluble ionic compound in water. The solubility product constant expression is the product of the concentrations of the ions with each concentration raised to a power equal to the coefficient of the ion in the chemical equation. For example, copper(II) hydroxide dissolves in water according to this equation.

$$Cu(OH)_2(s) \rightleftharpoons Cu^{2+}(aq) + 2OH^-(aq)$$

The coefficient of Cu^{2+} is 1, and the coefficient of OH^- is 2, so the following is the solubility product constant expression.

$$K_{sp} = [Cu^{2+}][OH^-]^2$$

Tabulated K_{sp} values may be used to calculate the molar solubility of a sparingly soluble ionic compound and also to calculate ion concentrations in a saturated solution. The following example problem illustrates these calculations.

Example Problem 18-5

Calculating Molar Solubility and Ion Concentration from K_{sp}

The K_{sp} for lead(II) fluoride (PbF_2) is 3.3×10^{-8} at 25°C. Use this K_{sp} value to calculate the following.

a. The solubility in mol/L of PbF_2

b. The fluoride ion concentration in a saturated solution of PbF_2

a. Write the balanced equation for the solubility equilibrium, and write the K_{sp} expression.

$$PbF_2(s) \rightleftharpoons Pb^{2+}(aq) + 2F^-(aq)$$

$$K_{sp} = [Pb^{2+}][F^-]^2 = 3.3 \times 10^{-8}$$

The moles of Pb^{2+} ions in solution equal the moles of PbF_2 that dissolved. Therefore, let $[Pb^{2+}]$ equal s, where s represents the solubility of PbF_2. Because there are two F^- ions for every Pb^{2+} ion, $[F^-] = 2s$. Substitute these terms into the K_{sp} expression and solve for s.

$$(s)(2s)^2 = 3.3 \times 10^{-8}$$

$$(s)(4s^2) = 3.3 \times 10^{-8}$$

$$4s^3 = 3.3 \times 10^{-8}$$

$$s^3 = \frac{3.3 \times 10^{-8}}{4} = 8.25 \times 10^{-9}$$

Here three digits are retained for accuracy, but the final answer will be rounded to two digits.

$$s = [Pb^{2+}] = \sqrt[3]{8.25 \times 10^{-9}} = 2.0 \times 10^{-3} \text{ mol/L}$$

The molar solubility of PbF_2 in water at 25°C is 2.0×10^{-3} mol/L.

b. As shown above, the fluoride ion concentration in a saturated solution of PbF_2 at 25°C is as follows.

$[F^-] = 2s = 2(2.0 \times 10^{-3} \text{ mol/L}) = 4.0 \times 10^{-3} \text{ mol/L}$

Practice Problems

12. Use the data in Table 18-3 in your textbook to calculate the solubility in mol/L of these ionic compounds at 298 K.
   a. $MgCO_3$
   b. $AlPO_4$
   c. Ag_2SO_4
   d. $Al(OH)_3$

13. Use the data in Table 18-3 in your textbook to calculate the following ion concentrations at 298 K.
   a. $[Ba^{2+}]$ in a saturated solution of $BaCrO_4$
   b. $[CO_3{}^{2-}]$ in a solution of $ZnCO_3$ at equilibrium
   c. $[Cl^-]$ in a solution of $PbCl_2$ at equilibrium
   d. $[Mg^{2+}]$ in a saturated solution of $Mg_3(PO_4)_2$

▶ **Predicting precipitates** The solubility product constant expression can also be used to predict whether a precipitate will form when two solutions of ionic compounds are mixed. The molar concentrations of the ions in a solution are used to calculate the ion product, Q_{sp}. If $Q_{sp} > K_{sp}$, a precipitate will form, reducing the ion concentrations until the system reaches equilibrium and the solution is saturated. If $Q_{sp} < K_{sp}$, no precipitate forms. The following example problem demonstrates how to use Q_{sp} and K_{sp} to determine whether a precipitate will form.

Example Problem 18-6
Predicting a Precipitate

Predict whether a precipitate will form if 200 mL of 0.030M $CaCl_2$ is added to 200 mL of 0.080M NaOH.

A double-replacement reaction might occur according to this equation.

$$CaCl_2 + 2NaOH \rightleftharpoons 2NaCl + Ca(OH)_2$$

You know that NaCl is a soluble compound and will not form a precipitate. However, $Ca(OH)_2$ is sparingly soluble with K_{sp} = 5.0×10^{-6}, so it might precipitate if the concentrations of its ions are high enough.

Write the equation for the dissolving of $Ca(OH)_2$.

$$Ca(OH)_2(s) \rightleftharpoons Ca^{2+}(aq) + 2OH^-(aq)$$

The ion product expression is as follows.

$$Q_{sp} = [Ca^{2+}][OH^-]^2$$

Q_{sp} is a trial value that will be compared to K_{sp}.

Next, find the concentrations of the Ca^{2+} and OH^- ions. Divide the initial concentrations in half because the volume doubles on mixing.

$$[Ca^{2+}] = \frac{0.030M}{2} = 0.015M$$

$$[OH^-] = \frac{0.080M}{2} = 0.040M$$

Calculate Q_{sp}.

$$Q_{sp} = (0.015)(0.040)^2 = 2.4 \times 10^{-5}$$

Compare Q_{sp} with K_{sp}.

$$Q_{sp} (2.4 \times 10^{-5}) > K_{sp} (5.0 \times 10^{-6})$$

The concentrations of Ca^{2+} and OH^- are high enough to cause a precipitate of $Ca(OH)_2$ to form.

Practice Problems

14. Use K_{sp} values from Table 18-3 in your textbook to predict whether a precipitate will form if equal volumes of these aqueous solutions are mixed.
  a. $0.010M$ $Ba(NO_3)_2$ and $0.050M$ Na_2SO_4
  b. $0.020M$ KBr and $0.015M$ $Pb(NO_3)_2$
  c. $0.0060M$ Na_2CrO_4 and $0.0025M$ $AgNO_3$

15. Will a precipitate form when 125 mL of $0.010M$ K_2SO_4 is mixed with 250 mL of $0.015M$ $CaBr_2$? (Hint: Note that the volumes of the two solutions are not equal, and the volume after mixing is 375 mL.)

▶ **Common ion effect** The solubility of a substance is reduced when the substance is dissolved in a solution containing a common ion. This is called the **common ion effect.** For example, PbI_2 is less soluble in an aqueous solution of NaI than in pure water because the common ion I^-, already present in the NaI solution, reduces the maximum possible concentration of Pb^{2+} and thus reduces the solubility of PbI_2.

Chapter 18 Review

16. If an equilibrium system contains small amounts of reactants and large amounts of products, what can you say about the value of K_{eq} for this equilibrium?

17. The following is the equation for a homogeneous equilibrium.

$$2A + B \rightleftharpoons 2C$$

What is the value of K_{eq} if [B] = 0.14 mol/L and [C] = 3.0 × [A]?

18. Is it possible to cause a shift in an equilibrium system without changing the equilibrium constant? Explain.

19. In a reversible endothermic reaction, four moles of gaseous reactants yield three moles of gaseous products. Describe four ways to shift this equilibrium toward the products.

20. At 298 K, the molar solubility of $Fe(OH)_2$ is higher than that of AgI, but K_{sp} for $Fe(OH)_2$ is less than K_{sp} for AgI. Explain how this is possible.

21. Would you expect Ag_2CO_3 to be more soluble in pure water or in a 1.0M Na_2CO_3 solution? Explain your answer.

Acids and Bases

19.1 Acids and Bases: An Introduction

All aqueous solutions contain hydrogen ions (H^+) and hydroxide ions (OH^-). An **acidic solution** contains more H^+ ions than OH^- ions, whereas a **basic solution** contains more OH^- ions than H^+ ions. A neutral solution contains equal concentrations of H^+ ions and OH^- ions. A hydronium ion (H_3O^+) is a hydrated hydrogen ion. The symbols H^+ and H_3O^+ can be used interchangeably in chemical equations.

In the **Arrhenius model** of acids and bases, an acid is a substance that contains hydrogen and ionizes in aqueous solution to produce hydrogen ions. A base is a substance that contains a hydroxide group and dissociates in aqueous solution to produce hydroxide ions. In the more inclusive **Brønsted-Lowry model,** an acid is a hydrogen-ion donor and a base is a hydrogen-ion acceptor. When a Brønsted-Lowry acid donates a hydrogen ion, a **conjugate base** is formed. When a Brønsted-Lowry base accepts a hydrogen ion, a **conjugate acid** is formed. Two substances related to each other by the donating and accepting of a single hydrogen ion are a **conjugate acid-base pair.**

Example Problem 19-1
Identifying Conjugate Acid-Base Pairs

Identify the conjugate acid-base pairs in this reaction.

$$HClO_2(aq) + H_2O(l) \rightleftharpoons H_3O^+(aq) + ClO_2^-(aq)$$

A hydrogen ion is donated by $HClO_2$, which is the Brønsted-Lowry acid in the forward reaction. The resulting conjugate base is ClO_2^-. The base in the forward reaction is H_2O, which accepts a hydrogen ion to form the conjugate acid H_3O^+.

Practice Problems

1. Identify the conjugate acid-base pairs in the following reactions.
   a. $H_2SO_3(aq) + H_2O(l) \rightleftharpoons HSO_3^-(aq) + H_3O^+(aq)$
   b. $HPO_4^{2-}(aq) + H_2O(l) \rightleftharpoons H_2PO_4^-(aq) + OH^-(aq)$
   c. $HSeO_3^-(aq) + H_2O(l) \rightleftharpoons H_3O^+(aq) + SeO_3^{2-}(aq)$

▶ **Monoprotic and polyprotic acids** An acid that can donate only one hydrogen ion is called a monoprotic acid. For example, hydrochloric acid (HCl) and formic acid (HCOOH) are monoprotic acids because they each contain only one ionizable hydrogen atom. Note that only those hydrogen atoms that are bonded to electronegative elements are ionizable.

Some acids can donate more than one hydrogen ion. For example, sulfuric acid (H_2SO_4) contains two ionizable hydrogen atoms, so it is called a diprotic acid. Boric acid (H_3BO_3) contains three ionizable hydrogen atoms, so it is a triprotic acid. More generally, an acid that contains two or more ionizable hydrogen atoms is called a polyprotic acid.

The complete ionization of a polyprotic acid occurs in steps. The three ionizations of boric acid are as follows.

$$H_3BO_3(aq) + H_2O(l) \rightleftharpoons H_3O^+(aq) + H_2BO_3^-(aq)$$

$$H_2BO_3^-(aq) + H_2O(l) \rightleftharpoons H_3O^+(aq) + HBO_3^{2-}(aq)$$

$$HBO_3^{2-}(aq) + H_2O(l) \rightleftharpoons H_3O^+(aq) + BO_3^{3-}(aq)$$

Practice Problems

2. Write the steps in the complete ionization of the following polyprotic acids.
   a. carbonic acid (H_2CO_3)
   b. chromic acid (H_2CrO_4)

19.2 Strengths of Acids and Bases

An acid that ionizes completely in dilute aqueous solution is called a **strong acid.** Examples include hydrochloric acid (HCl), nitric acid (HNO_3), sulfuric acid (H_2SO_4), and perchloric acid ($HClO_4$). A **weak acid** ionizes only partially in dilute aqueous solution. Some familiar examples of weak acids are carbonic acid (H_2CO_3), boric acid (H_3BO_3), phosphoric acid (H_3PO_4), and acetic acid ($HC_2H_3O_2$). The ionization of a weak acid reaches a state of equilibrium in which the forward and reverse reactions occur at equal rates. For example, consider the ionization equation for formic acid (HCOOH), a weak organic acid with numerous industrial uses.

$$HCOOH(aq) + H_2O(l) \rightleftharpoons H_3O^+(aq) + HCOO^-(aq)$$

The equilibrium constant expression for the ionization of formic acid in water is as follows.

$$K_a = \frac{[H_3O^+][HCOO^-]}{[HCOOH]}$$

K_a, called the **acid ionization constant,** is the value of the equilibrium constant expression for the ionization of a weak acid. The value of K_a indicates the extent of ionization of the acid. The weakest acids have the smallest K_a values. For formic acid, $K_a = 1.8 \times 10^{-4}$ at 298 K; this is considered a moderately weak acid.

In the case of a polyprotic acid, there is a K_a value for each ionization, and the K_a values decrease for each successive ionization. For example, the second ionization of phosphoric acid is represented by this equation.

$$H_2PO_4^-(aq) + H_2O(l) \rightleftharpoons H_3O^+(aq) + HPO_4^{2-}(aq)$$

The corresponding acid ionization constant expression is as follows.

$$K_a = \frac{[H_3O^+][HPO_4^{2-}]}{[H_2PO_4^-]}$$

Practice Problems

3. Write ionization equations and acid ionization constant expressions for the following acids.

    a. hydrofluoric acid (HF)

    b. hypobromous acid (HBrO)

4. Write the ionization equation and the acid ionization constant expression for the second ionization of sulfurous acid (H_2SO_3) in water.

▶ **Strengths of bases** Metallic hydroxides, such as potassium hydroxide, are **strong bases** because they dissociate entirely into metal ions and hydroxide ions in aqueous solution.

$$KOH(s) \rightarrow K^+(aq) + OH^-(aq)$$

A **weak base** ionizes only partially in dilute aqueous solution to form an equilibrium mixture. An example is the ionization of the weak base aniline ($C_6H_5NH_2$).

$$C_6H_5NH_2(aq) + H_2O(l) \rightleftharpoons C_6H_5NH_3^+(aq) + OH^-(aq)$$

The equilibrium constant expression for the ionization of aniline in water is as follows.

$$K_b = \frac{[C_6H_5NH_3^+][OH^-]}{[C_6H_5NH_2]}$$

K_b, called the **base ionization constant,** is the value of the equilibrium constant expression for the ionization of a weak base. As you might expect, K_b is smallest for the weakest bases. Aniline is a very weak base with $K_b = 4.3 \times 10^{-10}$ at 298 K.

Practice Problems

5. Write ionization equations and base ionization constant expressions for the following bases.

   a. butylamine ($C_4H_9NH_2$)

   b. phosphate ion (PO_4^{3-})

   c. hydrogen carbonate ion (HCO_3^-)

19.3 What is pH?

Pure water self-ionizes slightly to form H_3O^+ and OH^- ions, as shown in this equation.

$$H_2O(l) + H_2O(l) \rightleftharpoons H_3O^+(aq) + OH^-(aq)$$

The equation for the equilibrium can be simplified by removing one water molecule from each side.

$$H_2O(l) \rightleftharpoons H^+(aq) + OH^-(aq)$$

A special equilibrium expression for the self-ionization of water is defined as follows.

$$K_w = [H^+][OH^-]$$

K_w, called the **ion product constant for water,** is the value of the equilibrium constant expression for the self-ionization of water. In pure water at 298 K, the concentrations of H^+ ions and OH^- ions both equal $1.0 \times 10^{-7}M$, so the value of K_w can be calculated.

$$K_w = [H^+][OH^-] = (1.0 \times 10^{-7})(1.0 \times 10^{-7})$$

$$K_w = 1.0 \times 10^{-14}$$

At 298 K, the product of $[H^+]$ and $[OH^-]$ always equals 1.0×10^{-14}. Therefore, if the concentration of one of these ions increases, the concentration of the other ion must decrease. The following example problem shows how you can use K_w to find either $[H^+]$ or $[OH^-]$ if the other concentration is known.

Example Problem 19-2
Using K_w to Calculate $[H^+]$ and $[OH^-]$

At 298 K, the OH^- ion concentration of an aqueous solution is $1.0 \times 10^{-11}M$. Find the H^+ ion concentration in the solution and determine whether the solution is acidic, basic, or neutral.

Write the ion product constant expression.

$$K_w = [H^+][OH^-] = 1.0 \times 10^{-14}$$

Divide both sides of the equation by $[OH^-]$.

$$[H^+] = \frac{K_w}{[OH^-]}$$

Substitute the values for K_w and $[OH^-]$ and solve.

$$[H^+] = \frac{1.0 \times 10^{-14}}{1.0 \times 10^{-11}} = 1.0 \times 10^{-3}M$$

$[H^+] > [OH^-]$, so the solution is acidic.

Practice Problems

6. Given the concentration of either hydrogen ion or hydroxide ion, calculate the concentration of the other ion at 298 K and state whether the solution is acidic, basic, or neutral.

   a. $[OH^-] = 1.0 \times 10^{-6}M$

   b. $[H^+] = 1.0 \times 10^{-7}M$

   c. $[H^+] = 8.1 \times 10^{-3}M$

▶ **pH and pOH** Because the concentrations of H^+ ions are often very small numbers, the pH scale was developed as a more convenient way to express H^+ ion concentrations. The **pH** of a solution equals the negative logarithm of the hydrogen ion concentration.

$$pH = -\log [H^+]$$

The pH scale has values from 0 to 14. Acidic solutions have pH values between 0 and 7, with a value of 0 being the most acidic. The pH of a basic solution is between 7 and 14, with 14 representing the most basic solution. A neutral solution has a pH of 7.

Chemists have also defined a pOH scale to express the basicity of a solution. The **pOH** of a solution is the negative logarithm of the hydroxide ion concentration.

$$pOH = -\log [OH^-]$$

If either pH or pOH is known, the other may be determined by using the following relationship.

$$pH + pOH = 14.00$$

The pH and pOH values for a solution may be determined if either $[H^+]$ or $[OH^-]$ is known. The following example problem shows you how to calculate pH and pOH.

Example Problem 19-3
Calculating pH and pOH from [H⁺]

If a certain carbonated soft drink has a hydrogen ion concentration of $7.3 \times 10^{-4}M$, what are the pH and pOH of the soft drink?

Because $[H^+]$ is given, it is easier to calculate pH first.

$$pH = -\log [H^+]$$

$$pH = -\log [7.3 \times 10^{-4}]$$

$$pH = -(\log 7.3 + \log 10^{-4})$$

A log table or calculator shows that $\log 7.3 = 0.86$ and $\log 10^{-4} = -4$. Substitute these values in the equation for pH.

$$pH = -[0.86 + (-4)] = -(0.86 - 4) = 3.14$$

The pH of the soft drink is 3.14. Note that the number of decimal places retained in the pH value equals the number of significant figures in the H^+ ion concentration.

To find pOH, recall that $pH + pOH = 14.00$. Isolate pOH by subtracting pH from both sides of the equation.

$$pOH = 14.00 - pH$$

Substitute the value of pH and solve.

$$pOH = 14.00 - 3.14 = 10.86$$

The pOH of the solution is 10.86.

As you might expect, the carbonated soft drink is acidic.

Practice Problems

7. Calculate the pH and pOH of aqueous solutions having the following ion concentrations.

a. $[H^+] = 1.0 \times 10^{-14}M$

b. $[OH^-] = 5.6 \times 10^{-8}M$

c. $[H^+] = 2.7 \times 10^{-3}M$

d. $[OH^-] = 0.061M$

▶ **Calculating ion concentrations from pH** When the pH of a solution is known, you can determine the concentrations of H^+ and OH^-. First, recall the equation for pH.

$$pH = -\log [H^+]$$

Multiply both sides of the equation by -1.

$$-pH = \log [H^+]$$

Now take the antilog of both sides of the equation.

$$antilog (-pH) = [H^+]$$

Rearrange the equation.

$$[H^+] = antilog (-pH)$$

A similar relationship exists between $[OH^-]$ and pOH.

$$[OH^-] = antilog (-pOH)$$

Example Problem 19-4
Calculating $[H^+]$ and $[OH^-]$ from pH

What are $[H^+]$ and $[OH^-]$ in an antacid solution with a pH of 9.70?

Use pH to find $[H^+]$.

$$[H^+] = antilog (-pH)$$

$$[H^+] = antilog (-9.70)$$

Use a log table or calculator to find that the antilog of -9.70 is 2.0×10^{-10}.

$$[H^+] = 2.0 \times 10^{-10}M$$

To determine $[OH^-]$, first use the pH value to calculate pOH.

$$pOH = 14.00 - pH$$

$$pOH = 14.00 - 9.70 = 4.30$$

Now use the equation relating $[OH^-]$ to pOH.

$$[OH^-] = antilog\ (-pOH)$$

$$[OH^-] = antilog\ (-4.30)$$

A log table or calculator shows that the antilog of -4.30 is 5.0×10^{-5}.

$$[OH^-] = 5.0 \times 10^{-5}M$$

As expected, $[OH^-] > [H^+]$ in this basic solution.

Practice Problems

8. The pH or pOH is given for three solutions. Calculate $[H^+]$ and $[OH^-]$ in each solution.

   a. pH = 2.80

   b. pH = 13.19

   c. pOH = 8.76

▶ **Calculating the pH of strong acid and strong base solutions**
You learned in Section 19.2 that strong acids and strong bases ionize completely when dissolved in water. This means that for strong monoprotic acids, the concentration of the acid is the concentration of the H^+ ion because each acid molecule releases one H^+ ion. Similarly, for a strong base such as NaOH, the concentration of the base equals the concentration of the OH^- ion. However, some strong bases contain two or more hydroxide ions in each formula unit. An example is $Mg(OH)_2$. For a solution of $Mg(OH)_2$, $[OH^-]$ is twice the concentration of the base. For example, for a $3.0 \times 10^{-5}M$ $Mg(OH)_2$ solution, the concentration of OH^- is $2(3.0 \times 10^{-5}M) = 6.0 \times 10^{-5}M$. As you learned earlier in this section, pH can be calculated once $[H^+]$ or $[OH^-]$ is known.

Practice Problems

9. Calculate the pH of the following strong acid or strong base solutions.

a. $0.015M$ HCl

c. $2.5 \times 10^{-4}M$ HNO$_3$

b. $0.65M$ KOH

d. $4.0 \times 10^{-3}M$ Ca(OH)$_2$

▶ **Using pH to calculate K_a** If you know the pH and the concentration of a solution of a weak acid, you can calculate K_a for the acid. The following example problem illustrates this type of calculation.

Example Problem 19-5
Calculating K$_a$ from pH

The pH of a $0.200M$ solution of acetic acid (CH$_3$COOH) is 2.72. What is K_a for acetic acid?

Write the equation for the ionization reaction. For simplicity, water is omitted from the equation.

$$CH_3COOH(aq) \rightleftharpoons H^+(aq) + CH_3COO^-(aq)$$

The acid ionization constant expression is as follows.

$$K_a = \frac{[H^+][CH_3COO^-]}{[CH_3COOH]}$$

Use the pH to calculate [H$^+$].

$$pH = -\log [H^+]$$

$$[H^+] = \text{antilog} (-pH)$$

$$[H^+] = \text{antilog} (-2.72)$$

Use a log table or calculator to find that the antilog of -2.72 is 1.9×10^{-3}.

$$[H^+] = 1.9 \times 10^{-3}M$$

In the ionization reaction, equal numbers of H$^+$ ions and CH$_3$COO$^-$ ions are formed.

$$[CH_3COO^-] = [H^+] = 1.9 \times 10^{-3}M$$

At equilibrium, [CH$_3$COOH] equals the initial concentration of the acid minus the moles per liter that dissociated.

$$[CH_3COOH] = 0.200M - [H^+]$$

$$[CH_3COOH] = 0.200M - 1.9 \times 10^{-3}M = 0.198M$$

Substitute the known values into the equation for K_a and solve.

$$K_a = \frac{(1.9 \times 10^{-3})(1.9 \times 10^{-3})}{0.198} = 1.8 \times 10^{-5}$$

This K_a value indicates that CH_3COOH is a moderately weak acid.

Practice Problems

10. Calculate K_a for the following acids using the information provided.

  a. $0.100M$ solution of sulfurous acid (H_2SO_3), pH $= 1.48$

  b. $0.200M$ solution of benzoic acid (C_6H_5COOH), pH $= 2.45$

19.4 Neutralization

The reaction of an acid and a base in aqueous solution is called a **neutralization reaction.** The products of a neutralization reaction are a salt and water. A **salt** is an ionic compound composed of a positive ion from a base and a negative ion from an acid. An example of an acid-base neutralization is the reaction of nitric acid and calcium hydroxide to form calcium nitrate and water.

$$2HNO_3(aq) + Ca(OH)_2(aq) \rightarrow Ca(NO_3)_2(aq) + 2H_2O(l)$$

Acid-base neutralizations are used in the procedure called **titration,** which is a method for determining the concentration of a solution by reacting it with another solution of known concentration. For example, to find the concentration of an acid solution, you would slowly add a basic solution of known concentration. The neutralization reaction would proceed until it reaches the **equivalence point,** where the moles of H^+ ion from the acid equal the moles of OH^- ion from the base. At the equivalence point, a large change in pH occurs that can be detected by a pH meter or an **acid-base indicator,** which is a chemical dye whose color is affected by pH changes. The pH at the equivalence point depends upon the relative strengths of the acid and base used in the titration. For the reaction of a strong acid with a strong base, the pH at the equivalence point is 7. However, the pH at the equivalence point may be greater than 7 for a weak acid-strong base titration, or less than 7 for a strong acid-weak base titration. The following example problem shows you how to use titration data to determine the concentration of a solution.

Example Problem 19-6
Calculating Concentration from Titration Data

In a titration, 53.7 mL 0.100M HCl solution is needed to neutralize 80.0 mL of KOH solution. What is the molarity of the KOH solution?

Write the balanced equation for the neutralization reaction.

$$HCl(aq) + KOH(aq) \rightarrow KCl(aq) + H_2O(l)$$

Convert milliliters of HCl solution to liters.

$$53.7 \text{ mL HCl} \times \frac{1 \text{ L HCl}}{1000 \text{ mL HCl}} = 0.0537 \text{ L HCl}$$

Determine the moles of HCl used by multiplying the volume of the solution by its molarity, or mol/L.

$$0.0537 \text{ L HCl} \times \frac{0.100 \text{ mol HCl}}{1 \text{ L HCl}} = 5.37 \times 10^{-3} \text{ mol HCl}$$

Use the mole ratio in the balanced chemical equation to calculate the moles of KOH in the unknown solution.

$$5.37 \times 10^{-3} \text{ mol HCl} \times \frac{1 \text{ mol KOH}}{1 \text{ mol HCl}} = 5.37 \times 10^{-3} \text{ mol KOH}$$

Convert milliliters of KOH solution to liters.

$$80.0 \text{ mL KOH} \times \frac{1 \text{ L KOH}}{1000 \text{ mL KOH}} = 0.0800 \text{ L KOH}$$

Determine the molarity of the KOH solution by dividing the moles of KOH that reacted by the volume of the KOH solution in liters.

$$M_{KOH} = \frac{5.37 \times 10^{-3} \text{ mol KOH}}{0.0800 \text{ L KOH}} = 6.71 \times 10^{-2} M$$

The molarity of the KOH solution is $6.71 \times 10^{-2} M$, or $0.0671 M$.

Practice Problems

11. A 45.0-mL sample of nitric acid solution is neutralized by 119.4 mL 0.200M NaOH solution. What is the molarity of the nitric acid solution?

12. What is the molarity of a CsOH solution if 29.61 mL 0.2500M HCl is needed to neutralize 60.00 mL solution?

13. A 70.0-mL sample of sulfuric acid solution is neutralized by 256.3 mL 0.100M NaOH solution. What is the molarity of the sulfuric acid solution? (Hint: Note the mole ratio of the reactants in the balanced chemical equation.)

▶ **Buffered solutions** It is often desirable to reduce variations in pH when an acid or a base is added to a solution. A **buffer** is a solution that resists changes in pH when a moderate amount of acid or base is added. A buffer is a mixture of a weak acid and its conjugate base or a weak base and its conjugate acid. The buffer solution reacts with H^+ ions or OH^- ions added to it, thus maintaining a fairly constant pH value. An example of a buffer is the CH_3COOH/CH_3COO^- buffer system, which is made by mixing equal molar amounts of acetic acid (CH_3COOH) and an acetate salt such as potassium acetate (KCH_3COO) in water.

Chapter 19 Review

14. Explain how a base in the Arrhenius model is different from a base in the Brønsted-Lowry model.

15. The K_a value for acid X is 8.5×10^{-4}, while K_a for acid Y is 4.6×10^{-8}. Explain what these K_a values tell you about the strengths of the two acids.

16. How does the pH of an aqueous solution change when there is a decrease in the concentration of hydroxide ions? Explain your answer.

17. Solution A has a pH of 9.0, and solution B has a pOH of 3.0. State whether each solution is acidic, basic, or neutral. Which solution has a higher concentration of hydrogen ions?

18. What is the molarity of a solution of HCl if the pH of the solution is 2.00?

19. How can you detect the equivalence point of an acid-base titration?

20. A chemist prepares a buffer solution by dissolving sodium formate (NaHCOO) and another substance in water. What is the other substance likely to be? Explain your answer.

Redox Reactions

20.1 Oxidation and Reduction

A chemical reaction in which electrons are transferred from one atom to another is called an **oxidation–reduction reaction,** or **redox reaction.** For example, a thin sliver of zinc metal can be burned to form zinc oxide.

Complete chemical equation: $2Zn(s) + O_2(g) \rightarrow 2ZnO(s)$

Net ionic equation: $2Zn(s) + O_2(g) \rightarrow 2Zn^{2+} + 2O^{2-}$ (ions in crystal)

In this reaction, each zinc atom transfers two electrons to an oxygen atom. The zinc atoms become Zn^{2+} ions, while the oxygen atoms become O^{2-} ions.

In a redox reaction, the loss of electrons from atoms of a substance is called **oxidation,** whereas the gain of electrons is called **reduction.** In the reaction shown above, zinc loses electrons and is therefore oxidized. Oxygen gains electrons and is therefore reduced. You learned in previous chapters that the oxidation number of an atom in an ionic compound equals the number of electrons lost or gained by the atom when it forms an ion. Oxidation increases an atom's oxidation number; reduction decreases the oxidation number. In zinc oxide, the oxidation number of zinc is $+2$, and the oxidation number of oxygen is -2. Note that oxidation numbers are written with the positive or negative sign before the number ($+2$, -2). Ionic charge is written with the sign after the number ($2+$, $2-$).

Oxidation and reduction are complementary processes that always occur together. The substance that is reduced in a redox reaction is called the **oxidizing agent.** The substance that is oxidized is called the **reducing agent.** In the formation of zinc oxide, oxygen is the oxidizing agent and zinc is the reducing agent.

Redox reactions are not limited to reactions in which atoms change to ions or vice versa. For example, consider the synthesis of hydrogen chloride gas from its elements.

$$H_2(g) + Cl_2(g) \rightarrow 2HCl(g)$$

This reaction is a redox reaction even though the product is a molecular compound and no ions are involved in the process. The more electronegative atom (chlorine) is considered to have been reduced by gaining electrons, and the less electronegative atom (hydrogen) is considered to have been oxidized by losing electrons. Therefore, in this reaction, chlorine is the oxidizing agent and hydrogen is the reducing agent. The electronegativity values in Figure 6-18 in your textbook may be helpful.

Practice Problems

1. For each of the following reactions, identify what is oxidized and what is reduced. Also identify the oxidizing agent and the reducing agent.

a. $Zn + Ni^{2+} \rightarrow Ni + Zn^{2+}$ **c.** $2NO \rightarrow N_2 + O_2$

b. $2I^- + Br_2 \rightarrow I_2 + 2Br^-$ **d.** $2H_2 + S_2 \rightarrow 2H_2S$

▶ **Determining oxidation numbers** When a redox reaction occurs, there is a change in the oxidation number of each atom that is oxidized or reduced. In order to understand these changes, you must be able to determine the oxidation number of an element in a compound. Chemists have developed the following rules for determining oxidation numbers.

1. *The oxidation number of an uncombined atom is zero.* Therefore, free elements have an oxidation number of zero.

2. *The oxidation number of a monatomic ion is equal to the charge on the ion.* For example, the oxidation number of a Fe^{3+} ion is $+3$, and the oxidation number of a Cl^- ion is -1.

3. *The oxidation number of the more electronegative atom in a molecule or a complex ion is the same as the charge it would have if it were an ion.* For example, in the compound phosphorus pentabromide (PBr_5), bromine is more electronegative than phosphorus. Therefore, each bromine is given an oxidation number of -1, as if it had gained an electron to complete an octet. Phosphorus is assigned an oxidation number of $+5$, as if it had lost an electron to each bromine atom.

4. *The most electronegative element, fluorine, always has an oxidation number of -1 when it is bonded to another element.*

5. *The oxidation number of oxygen in compounds is always* -2, *except in peroxides, such as hydrogen peroxide* (H_2O_2), *where it is* -1. *When it is bonded to fluorine, the oxidation number of oxygen is* $+2$.

6. *The oxidation number of hydrogen in most of its compounds is* $+1$. The exception to this rule is the metal hydrides, such as LiH, MgH_2, and AlH_3. In these compounds, hydrogen has an oxidation number of -1.

7. *The metals of groups 1A and 2A and aluminum in group 3A form compounds in which the metal atom always has a positive oxidation number equal to the number of its valence electrons* $(+1, +2, and +3, respectively)$.

8. *The sum of the oxidation numbers in a neutral compound is zero.*

9. *The sum of the oxidation numbers of the atoms in a polyatomic ion is equal to the charge on the ion.*

Many elements not specified in the rules above can have different oxidation numbers in different compounds. The list of rules enables you to determine these unknown oxidation numbers, as the following example problem demonstrates.

Example Problem 20-1
Determining Oxidation Numbers

Determine the oxidation number of each element in the following compound and ion.

   a. $SrCO_3$ (strontium carbonate)   **b.** $Cr_2O_7{}^{2-}$ (dichromate ion)

Assign the known oxidation numbers to their elements, set the sum of all oxidation numbers to zero or to the ion charge, and solve for the unknown oxidation number ($n_{element}$).

   a. According to rule 8, the sum of the oxidation numbers is zero because strontium carbonate is a neutral compound. Rule 5 states that the oxidation number of oxygen in compounds is -2, and rule 7 states that group 2 metals, such as strontium, have a $+2$ oxidation number in compounds.

$$(+2) + (n_C) + 3(-2) = 0$$
$$SrCO_3$$
$$2 + n_C + (-6) = 0$$
$$n_C = +4$$

The oxidation number of carbon is $+4$.

b. The dichromate ion has a charge of $2-$, so rule 9 says that the oxidation numbers add up to -2. According to rule 5, the oxidation number of oxygen in compounds is -2.

$$2(n_{Cr}) + 7(-2) = -2$$
$$Cr_2O_7^{2-}$$
$$2(n_{Cr}) + (-14) = -2$$
$$2(n_{Cr}) = +12$$
$$n_{Cr} = +6$$

The oxidation number of chromium is $+6$.

Practice Problems

2. Determine the oxidation number of the boldface element in each of these compounds.

   a. Li_2SiO_3 **c.** CaH_2 **e.** K_2GeF_6

   b. Al_4C_3 **d.** $BeSeO_4$ **f.** $Al(ClO_3)_3$

3. Determine the oxidation number of the boldface element in each of these ions.

   a. PO_4^{3-} **c.** HSO_4^- **e.** PuO_2^+

   b. Hg_2^{2+} **d.** $PtCl_6^{2-}$ **f.** TeO_3^{2-}

Not all chemical reactions can be classified as redox reactions. For example, in a double-replacement reaction, the positive and negative ions of two compounds are interchanged. Most double-replacement reactions are not redox reactions because there is no transfer of electrons between atoms. Combustion and single-replacement reactions, however, are always redox reactions. Many synthesis and decomposition reactions are redox reactions as well.

20.2 Balancing Redox Equations

Chemists use a technique for balancing redox equations that is based on the fact that the total increase in oxidation numbers resulting from oxidation must equal the total decrease in oxidation numbers resulting from reduction. The technique, called the **oxidation-number method,** consists of five steps.

Step 1 Assign oxidation numbers to all atoms in the equation.

Step 2 Identify the atoms that are oxidized and the atoms that are reduced.

Step 3 Determine the change in oxidation number for the atoms that are oxidized and for the atoms that are reduced.

Step 4 Make the changes in oxidation number equal in magnitude by adjusting coefficients in the equation.

Step 5 If necessary, use the conventional method to balance the remainder of the equation.

The following example problem shows you how to use the oxidation-number method to balance a redox equation.

Example Problem 20-2
Balancing a Redox Equation by the Oxidation-Number Method

The reaction of antimony with hot sulfuric acid produces antimony(III) sulfate, sulfur dioxide gas, and water, as shown in this unbalanced equation.

$$Sb(s) + H_2SO_4(aq) \rightarrow Sb_2(SO_4)_3(aq) + SO_2(g) + H_2O(l)$$

Balance this redox equation using the oxidation-number method.

Step 1 Assign oxidation numbers to all of the atoms in the equation, using the rules in Section 20.1.

$$\overset{0}{Sb} + \overset{+1\ +6\ -2}{H_2SO_4} \rightarrow \overset{+3\ +6\ -2}{Sb_2(SO_4)_3} + \overset{+4\ -2}{SO_2} + \overset{+1\ -2}{H_2O}$$

Step 2 Identify which atoms are oxidized and which are reduced.

$$\overset{0}{Sb} + \overset{+1\ +6\ -2}{H_2SO_4} \rightarrow \overset{+3\ +6\ -2}{Sb_2(SO_4)_3} + \overset{+4\ -2}{SO_2} + \overset{+1\ -2}{H_2O}$$

Antimony is oxidized in the reaction, while sulfur is reduced in the
formation of SO_2. The oxidation number of antimony increases from
0 to +3, and the oxidation number of sulfur decreases from +6 to
+4. The oxidation numbers of hydrogen and oxygen are unchanged.
The sulfate ion (SO_4^{2-}) appears on both sides of the equation, and
its atoms are neither oxidized nor reduced.

Step 3 Draw a line connecting the atoms involved in oxidation and
another line connecting the atoms involved in reduction. Write the
change in oxidation number corresponding to each line.

$$\overset{+3}{Sb} + H_2SO_4 \rightarrow Sb_2(SO_4)_3 + SO_2 + H_2O$$
$$-2$$

Step 4 Make the changes in oxidation number equal in magnitude
by placing the appropriate coefficients in the equation. The oxidation number change for antimony is +3, and the oxidation number
change for sulfur is −2. The magnitudes can be made equal by
adding a coefficient of 2 to antimony and adding a coefficient of 3 to
sulfur in the chemical equation. The coefficient of 3 is added to
H_2SO_4 on the left side of the equation and to SO_2 on the right side.
For antimony, note that two atoms of Sb are already present on the
right side of the equation. Therefore, the coefficient of 2 is added
only to Sb on the left side.

$$2(+3) = +6$$
$$2Sb + 3H_2SO_4 \rightarrow Sb_2(SO_4)_3 + 3SO_2 + H_2O$$
$$3(-2) = -6$$

Step 5 Balance the remainder of the equation by using the conventional method.

$$2Sb + 3H_2SO_4 \rightarrow Sb_2(SO_4)_3 + 3SO_2 + H_2O$$

Increase the coefficient of H_2SO_4 to 6 to balance the six sulfur
atoms on the right. (Note that only three sulfur atoms are reduced.)

$$2Sb + 6H_2SO_4 \rightarrow Sb_2(SO_4)_3 + 3SO_2 + H_2O$$

Add a coefficient of 6 to H_2O to balance the 12 hydrogen atoms on the left. This also balances the oxygen atoms, with 24 on each side.

$$2Sb(s) + 6H_2SO_4(aq) \rightarrow Sb_2(SO_4)_3(aq) + 3SO_2(g) + 6H_2O(l)$$

The equation is now balanced.

Practice Problems

4. Use the oxidation-number method to balance these redox equations.

a. $Cu_2O + NO \rightarrow CuO + N_2$

b. $Al_2O_3 + C + N_2 \rightarrow AlN + CO$

c. $Ag + HNO_3 \rightarrow AgNO_3 + NO + H_2O$

d. $I_2 + HClO + H_2O \rightarrow HIO_3 + HCl$

▶ **Balancing net ionic redox equations** The simplest way to express a redox reaction is an equation that shows only the oxidation and reduction processes. In order to understand how this is done, consider the balanced equation for the reaction of iron(II) nitrate and nitric acid.

$$3Fe(NO_3)_2(aq) + 4HNO_3(aq) \rightarrow$$
$$3Fe(NO_3)_3(aq) + NO(g) + 2H_2O(l)$$

You can confirm that iron is oxidized in the reaction, and nitrogen is reduced in the formation of nitrogen monoxide gas.

In Chapter 10, you learned how to write net ionic equations for chemical reactions. For the reaction shown above, the balanced net ionic equation is as follows.

$$3Fe^{2+}(aq) + 4H^+(aq) + NO_3^-(aq) \rightarrow$$
$$3Fe^{3+}(aq) + NO(g) + 2H_2O(l)$$

This equation can also be presented in unbalanced form.

$$Fe^{2+}(aq) + H^+(aq) + NO_3^-(aq) \rightarrow Fe^{3+}(aq) + NO(g) + H_2O(l)$$

Finally, the equation can be written to show only the substances that are oxidized and reduced. The hydrogen ion (H^+) and the water molecule are neither oxidized nor reduced, so they are removed from the equation.

$$Fe^{2+}(aq) + NO_3^-(aq) \rightarrow Fe^{3+}(aq) + NO(g) \text{ (in acid solution)}$$

Note that this equation shows that the reaction occurs in acid solution. This is important because in acid solution, hydrogen ions and water molecules are abundant and able to participate as either reactants or products in redox reactions. When a redox reaction occurs in basic solution, hydroxide ions (OH^-) and water molecules are abundant and free to react.

The following example problem shows you how to balance a net ionic equation for a redox reaction using the oxidation-number method.

Example Problem 20-3
Balancing a Net Ionic Redox Equation

Use the oxidation-number method to balance the net ionic equation for the redox reaction between the permanganate ion and the chloride ion in acid solution.

$$MnO_4^-(aq) + Cl^-(aq) \rightarrow Mn^{2+}(aq) + Cl_2(g) \text{ (in acid solution)}$$

Step 1 Use the rules in Section 20.1 to assign oxidation numbers to all atoms in the equation.

$$\overset{+7\ -2}{MnO_4^-}(aq) + \overset{-1}{Cl^-}(aq) \rightarrow \overset{+2}{Mn^{2+}}(aq) + \overset{0}{Cl_2}(g) \text{ (in acid solution)}$$

Step 2 Identify which atoms are oxidized and which are reduced.

$$\overset{+7\ -2}{MnO_4^-}(aq) + \overset{-1}{Cl^-}(aq) \rightarrow \overset{+2}{Mn^{2+}}(aq) + \overset{0}{Cl_2}(g) \text{ (in acid solution)}$$

The oxidation number of chlorine increases from -1 to 0 as it is oxidized in the reaction. The oxidation number of manganese decreases from $+7$ to $+2$ as it is reduced. No oxygen atoms appear in the products of the net ionic equation; they will be added later.

Step 3 Draw a line connecting the atoms involved in oxidation and another line connecting the atoms involved in reduction. Write the change in oxidation number corresponding to each line.

$$\overset{+1}{\overbrace{MnO_4^-(aq) + Cl^-(aq) \rightarrow Mn^{2+}(aq) + Cl_2(g)}} \text{ (in acid solution)}$$

$$\underbrace{\qquad\qquad\qquad}_{-5}$$

Step 4 Make the changes in oxidation number equal in magnitude by placing the appropriate coefficients in the equation. The oxidation number change for chlorine is +1, and the oxidation number change for manganese is −5. Normally, you would add a coefficient of 5 to chlorine in the equation to make the magnitudes equal. However, note that chlorine atoms only appear in even numbers in the products, so the equation must contain an even number of chlorine atoms. You can accomplish this by doubling the coefficient to 10 for chlorine, and also doubling the coefficient for manganese, so that the changes in oxidation number are balanced.

$$10(+1) = +10$$

$$2MnO_4^-(aq) + 10Cl^-(aq) \rightarrow 2Mn^{2+}(aq) + 5Cl_2(g) \text{ (in acid solution)}$$

$$2(-5) = -10$$

Note that $5Cl_2$ represents 10 chlorine atoms in the products, so chlorine is balanced in the equation.

Step 5 The reaction occurs in acid solution. To balance the equation, add enough water molecules to the equation to balance the oxygen atoms on both sides of the equation. Then add enough hydrogen ions to balance hydrogen on both sides.

$$2MnO_4^-(aq) + 10Cl^-(aq) + 16H^+(aq) \rightarrow$$
$$2Mn^{2+}(aq) + 5Cl_2(g) + 8H_2O(l)$$

The atoms and charges are now balanced.

Practice Problems

5. Use the oxidation-number method to balance these net ionic redox equations.

   a. $Al(s) + Ni^{2+}(aq) \rightarrow Al^{3+}(aq) + Ni(s)$

   b. $HS^-(aq) + IO_3^-(aq) \rightarrow I^-(aq) + S(s)$ (in acid solution)

c. $I_2(s) + HClO(aq) \rightarrow IO_3^-(aq) + Cl^-(aq)$ (in acid solution)

d. $MnO_4^{2-}(aq) \rightarrow MnO_4^-(aq) + MnO_2(s)$ (in acid solution)

20.3 Half-Reactions

The oxidation process and the reduction process of a redox reaction can each be expressed as a **half-reaction.** For example, consider the unbalanced equation for the formation of aluminum bromide.

$$Al + Br_2 \rightarrow AlBr_3$$

The oxidation half-reaction shows the loss of electrons by aluminum.

$$Al \rightarrow Al^{3+} + 3e^-$$

The reduction half-reaction shows the gain of electrons by bromine.

$$Br_2 + 2e^- \rightarrow 2Br^-$$

You can use half-reactions to balance a redox equation by following these five steps.

Step 1 Write the net ionic equation for the reaction, omitting spectator ions.

Step 2 Write the oxidation and reduction half-reactions for the net ionic equation.

Step 3 Balance the atoms and charges in each half-reaction.

Step 4 Adjust the coefficients so that the number of electrons lost in oxidation equals the number of electrons gained in reduction.

Step 5 Add the balanced half-reactions and return spectator ions.

Example Problem 20-4
Balancing a Redox Equation by Half-Reactions

Use the half-reaction method to balance this redox equation.

$$K_2Cr_2O_7(aq) + HCl(aq) \rightarrow CrCl_3(aq) + KCl(aq) + Cl_2(g)$$

Step 1 The strong acid HCl is a reactant, and the reaction occurs in acid solution. The four dissolved compounds exist as ions in solution. Some of the ions are neither oxidized nor reduced in the reaction. These ions can be removed from the equation. In this reaction, the ions to be removed are the potassium and hydrogen ions, as well as the chloride ions in the products. Eliminating these ions yields the net ionic equation.

$$Cr_2O_7^{2-}(aq) + Cl^-(aq) \rightarrow Cr^{3+}(aq) + Cl_2(g)$$

Step 2 Write the oxidation and reduction half-reactions, including oxidation numbers.

$$\overset{-1}{Cl^-} \rightarrow \overset{0}{Cl_2} + e^- \text{ (oxidation)}$$

$$\overset{+6}{Cr_2O_7^{2-}} + 3e^- \rightarrow \overset{+3}{Cr^{3+}} \text{ (reduction)}$$

Step 3 Balance the atoms and charges in each half-reaction. The oxidation half-reaction is easily balanced.

$$2Cl^- \rightarrow Cl_2 + 2e^- \text{ (oxidation)}$$

For the reduction half-reaction, first balance chromium by adding a Cr^{3+} ion to the right side of the equation. Six electrons are gained by the two chromium atoms during reduction.

$$Cr_2O_7^{2-} + 6e^- \rightarrow 2Cr^{3+}$$

The reaction occurs in acid solution. Add water molecules to the right side of the equation to balance oxygen. Then add H^+ ions to the left side to balance hydrogen atoms and charges.

$$Cr_2O_7^{2-} + 6e^- + 14H^+ \rightarrow 2Cr^{3+} + 7H_2O \text{ (reduction)}$$

Step 4 Adjust the coefficients so that the number of electrons lost in oxidation (2) equals the number of electrons gained in reduction (6). To do this, multiply the oxidation half-reaction by 3.

$$6Cl^- \rightarrow 3Cl_2 + 6e^- \text{ (oxidation)}$$

$$Cr_2O_7^{2-} + 6e^- + 14H^+ \rightarrow 2Cr^{3+} + 7H_2O \text{ (reduction)}$$

Step 5 Add the balanced half-reactions and cancel like terms on both sides of the equation.

$$Cr_2O_7^{2-} + 6Cl^- + 14H^+ \rightarrow 2Cr^{3+} + 3Cl_2 + 7H_2O$$

Return the spectator ions (K^+ and Cl^-). Two K^+ ions go with the $Cr_2O_7^{2-}$ ion on the left. On the right, six Cl^- ions join with the two Cr^{3+} ions, and two Cl^- ions combine with the two restored K^+

ions. To maintain balance, you must also add eight Cl^- ions to the left, where they join with the hydrogen ions.

$$K_2Cr_2O_7(aq) + 14HCl(aq) \rightarrow$$
$$2CrCl_3(aq) + 2KCl(aq) + 3Cl_2(g) + 7H_2O(l)$$

The chemical equation is balanced, with the state descriptions restored. Note that no subscripts have been changed.

Practice Problems

6. Use the half-reaction method to balance these redox equations. Start with step 2 of Example Problem 20-4 and leave the balanced equation in ionic form.
   a. $I_2(s) + H_2SO_3(aq) \rightarrow I^-(aq) + HSO_4^-(aq)$
   b. $Fe^{2+}(aq) + MnO_4^-(aq) \rightarrow Fe^{3+}(aq) + Mn^{2+}(aq)$
   c. $Zn(s) + Cr_2O_7^{2-}(aq) \rightarrow Zn^{2+}(aq) + Cr^{3+}(aq)$
   d. $IO_3^-(aq) + I^-(aq) \rightarrow I_2(s)$

Chapter 20 Review

7. Which of these unbalanced equations do NOT represent redox reactions? Explain your answer.
   a. $C_2H_6 + O_2 \rightarrow CO_2 + H_2O$
   b. $BaCl_2 + NaIO_3 \rightarrow Ba(IO_3)_2 + NaCl$
   c. $Al(NO_3)_3 + KOH \rightarrow Al(OH)_3 + KNO_3$
   d. $Hg_2O \rightarrow HgO + Hg$

8. Determine the oxidation number of bromine in each of the following compounds. Explain your answers.
   a. LiBr   b. BrF_3 c. $Mg(BrO_3)_2$

9. Which element is a stronger oxidizing agent—iodine or chlorine? Explain.

10. Write a balanced equation for the high-temperature reaction of iron(III) oxide and carbon monoxide gas to form molten iron metal and carbon dioxide gas.

11. Balance the following redox equation by using the half-reaction method. Explain the steps you followed. Give the final equation as it is shown below but with the balancing coefficients.
    $$HClO_3(aq) \rightarrow HClO_4(aq) + ClO_2(g) + H_2O(l)$$

Electrochemistry

21.1 Voltaic Cells

Electrochemistry is the study of processes by which chemical energy is converted to electrical energy and vice versa. This branch of chemistry has numerous applications in today's increasingly technological world, such as for batteries, electronic components, and commercial electroplating.

In Chapter 20, you learned that electrons are transferred between atoms in oxidation–reduction (redox) reactions. Consider the redox reaction represented by the following net ionic equation.

$$Ni(s) + Cu^{2+}(aq) \rightarrow Ni^{2+}(aq) + Cu(s)$$

In this reaction, nickel metal is oxidized to Ni^{2+} ions, and Cu^{2+} ions are reduced to copper metal. The half-reactions are as follows.

$$Ni(s) \rightarrow Ni^{2+}(aq) + 2e^- \text{ (oxidation)}$$

$$Cu^{2+}(aq) + 2e^- \rightarrow Cu(s) \text{ (reduction)}$$

This redox reaction can be used to generate an electric current by using an **electrochemical cell,** an apparatus that uses a redox reaction to produce electrical energy (**voltaic cell**) or uses electrical energy to cause a chemical reaction (**electrolytic cell**). The voltaic cell for this reaction begins with a strip of nickel metal in an aqueous solution of Ni^{2+} ions and a strip of copper metal in a solution of Cu^{2+} ions. Then, a metal-conducting wire is attached to the metal strips in the solutions. Finally, a **salt bridge** connects the two solutions. A salt bridge is usually a tube containing a conducting solution of a soluble salt, held in place by a plug, such as cotton or agar gel. The salt bridge allows the passage of ions into the two solutions but prevents the solutions from mixing.

When the voltaic cell is complete, the redox reaction proceeds. Electrons flow through the wire from the nickel strip to the copper strip, creating an electric current. The nickel strip is oxidized to Ni^{2+} ions, while the Cu^{2+} ions in the other solution are reduced to copper metal, which plates out on the copper strip. Ions flow from the salt bridge into the two solutions to maintain the balance of charges.

The two parts of a voltaic cell are called **half-cells.** Each half-cell contains an electrode, which conducts electrons into or out of the half-cell. The electrode where oxidation occurs is called the **anode,** and the electrode where reduction occurs is called the **cathode.** In the previous example, the nickel strip is the anode and the copper strip is the cathode.

▶ **Calculating cell potential** The tendency of a substance to gain electrons is called its **reduction potential.** The reduction potential of a half-cell reaction is expressed in volts (V). Chemists measure the reduction potentials of half-cells against the **standard hydrogen electrode,** which has a reduction potential defined as 0 V at 25°C, 1.0 atm pressure, and $1M$ hydrogen ion concentration.

Table 21-1 on page 667 of your textbook lists the standard reduction potentials (E^0 values) of common half-cell reactions. These values apply to the standard conditions of a $1M$ solution, 25°C, and 1 atm pressure. In any voltaic cell, the half-reaction with the lower reduction potential will proceed in the opposite direction, as an oxidation half-reaction. The half-reaction with the higher reduction potential will proceed as a reduction. The electrical potential of a voltaic cell, also called the cell potential, is found by subtracting the standard reduction potential of the oxidation half-reaction from the standard reduction potential of the reduction half-reaction.

$$E^0_{cell} = E^0_{reduction} - E^0_{oxidation}$$

For example, suppose the redox reaction considered previously is carried out in a voltaic cell under standard conditions. The E^0 values of the half-reactions can be found in Table 21-1 in your textbook.

$$Ni^{2+}(aq) + 2e^- \rightarrow Ni(s) \quad E^0 = -0.257 \text{ V}$$

$$Cu^{2+}(aq) + 2e^- \rightarrow Cu(s) \quad E^0 = +0.3419 \text{ V}$$

The first half-reaction has the lower reduction potential, so it proceeds as an oxidation. As expected, the half-reactions of the overall redox reaction are as follows.

$$Ni(s) \rightarrow Ni^{2+}(aq) + 2e^- \text{ (oxidation)}$$

$$Cu^{2+}(aq) + 2e^- \rightarrow Cu(s) \text{ (reduction)}$$

The cell potential is found as follows.

$$E^0_{cell} = E^0_{reduction} - E^0_{oxidation}$$

$$= +0.3419 \text{ V} - (-0.257 \text{ V})$$

$$= +0.599 \text{ V}$$

The overall reaction can be conveniently expressed in a form called cell notation.

$$Ni|Ni^{2+}\|Cu^{2+}|Cu$$

In cell notation, the oxidation reactant and product appear on the left, followed by two vertical lines, and the reduction reactant and product appear on the right.

The following example problem illustrates how to calculate the potential of a voltaic cell.

Example Problem 21-1
Calculating Cell Potential

The half-cells of a voltaic cell are represented by these two reduction half-reactions.

$$Cr^{2+}(aq) + 2e^- \rightarrow Cr(s)$$

$$Al^{3+}(aq) + 3e^- \rightarrow Al(s)$$

Determine the overall cell reaction and the standard cell potential. Express the reaction using cell notation.

The standard reduction potentials are found in Table 21-1 in your textbook.

$$Cr^{2+}(aq) + 2e^- \rightarrow Cr(s) \quad E^0_{Cr^{2+}|Cr} = -0.913 \text{ V}$$

$$Al^{3+}(aq) + 3e^- \rightarrow Al(s) \quad E^0_{Al^{3+}|Al} = -1.662 \text{ V}$$

The reduction of chromium has the higher (less negative) reduction potential, so this half-reaction proceeds as a reduction. The aluminum half-reaction proceeds in the opposite direction as an oxidation.

$$Cr^{2+}(aq) + 2e^- \rightarrow Cr(s) \text{ (reduction half-reaction)}$$

$$Al(s) \rightarrow Al^{3+}(aq) + 3e^- \text{ (oxidation half-reaction)}$$

Balance electrons in the half-reactions by multiplying the reduction half-reaction by 3 and the oxidation half-reaction by 2. Then add the half-reactions to obtain the overall cell reaction.

$$3Cr^{2+}(aq) + 6e^- \rightarrow 3Cr(s)$$
$$2Al(s) \rightarrow 2Al^{3+}(aq) + 6e^-$$
$$\overline{3Cr^{2+}(aq) + 2Al(s) \rightarrow 3Cr(s) + 2Al^{3+}(aq)}$$

Calculate the standard potential of the voltaic cell.

$$E^0_{cell} = E^0_{reduction} - E^0_{oxidation}$$

$$E^0_{cell} = E^0_{Cr^{2+}|Cr} - E^0_{Al^{3+}|Al}$$

$$= -0.913 \text{ V} - (-1.662 \text{ V})$$

$$= +0.749 \text{ V}$$

Notice that the standard reduction potentials of the half-reactions are *not* multiplied by their coefficients in the balanced overall equation. Recall your study of properties of matter from Chapter 3. Standard potential is an intensive property that does not depend on the amount of material used.

The overall reaction can be expressed in cell notation, with oxidation on the left and reduction on the right.

$$Al|Al^{3+}||Cr^{2+}|Cr$$

Practice Problems

1. For each pair of half-reactions, write the balanced equation for the overall cell reaction, calculate the standard cell potential, and express the reaction using cell notation. Use E^0 values from Table 21-1 in your textbook.

   a. $Mg^{2+}(aq) + 2e^- \rightarrow Mg(s)$
   $Pd^{2+}(aq) + 2e^- \rightarrow Pd(s)$

   b. $Cu^+(aq) + e^- \rightarrow Cu(s)$
   $Cd^{2+}(aq) + 2e^- \rightarrow Cd(s)$

   c. $Ce^{3+}(aq) + 3e^- \rightarrow Ce(s)$
   $2H^+(aq) + 2e^- \rightarrow H_2(g)$

2. Calculate the standard cell potentials of voltaic cells that contain the following pairs of half-cells. Use E^0 values from Table 21-1 in your textbook.

   a. iron in a solution of Fe^{2+} ions; silver in a solution of Ag^+ ions

   b. chlorine in a solution of Cl^- ions; zinc in a solution of Zn^{2+} ions

▶ **Determining reaction spontaneity** An important use of standard reduction potentials is to determine if a proposed reaction is spontaneous under standard conditions. To predict whether a proposed redox reaction is spontaneous, write the reaction in the form of half-reactions and look up the standard reduction potentials. Use the values to calculate the standard potential of a voltaic cell with the two half-cell reactions. If the calculated voltage is positive, the reaction is spontaneous. If the value is negative, the reaction is not spontaneous; however, the reverse reaction would be spontaneous because it would have a positive cell voltage.

Example Problem 21-2
Predicting the Spontaneity of a Reaction

Predict whether the following redox reaction will occur spontaneously.

$$2Cr^{3+}(aq) + 3Sn^{2+}(aq) \rightarrow 2Cr(s) + 3Sn^{4+}(aq)$$

Write the half-reactions. Note that the coefficients are simplified.

$$Cr^{3+}(aq) + 3e^- \rightarrow Cr(s) \text{ (reduction)}$$

$$Sn^{2+}(aq) \rightarrow Sn^{4+}(aq) + 2e^- \text{ (oxidation)}$$

Find the standard cell potential, using E^0 values from Table 21-1 in your textbook.

$$E^0_{cell} = E^0_{Cr^{3+}|Cr} - E^0_{Sn^{4+}|Sn^{2+}}$$

$$= -0.744V - (+0.151 \text{ V})$$

$$= -0.895 \text{ V}$$

The voltage is negative, so the reaction is not spontaneous. The reverse reaction will occur spontaneously.

Practice Problems

3. Calculate the cell potential to determine if each of these redox reactions is spontaneous. The half-reactions can be found in Table 21-1 in your textbook.

a. $2Ag^+(aq) + Co(s) \rightarrow Co^{2+}(aq) + 2Ag(s)$

b. $Cu(s) + Cu^{2+}(aq) \rightarrow 2Cu^+(aq)$

c. $2Br_2(l) + 2H_2O(l) \rightarrow 4H^+(aq) + 4Br^-(aq) + O_2(g)$

d. $5Fe^{2+}(aq) + MnO_4^-(aq) + 8H^+(aq) \rightarrow$
$$5Fe^{3+}(aq) + Mn^{2+}(aq) + 4H_2O(l)$$

21.2 Types of Batteries

A **battery** is one or more voltaic cells in a single package that generates an electric current. There are several types of batteries with various applications.

▶ **Dry cells** You are probably most familiar with the type of battery knows as the **dry cell,** which contains a moist paste in which the cathode half-reaction occurs. Three types of dry cells are the zinc–carbon dry cell, the alkaline cell, and the mercury cell. Zinc is oxidized in each of these dry cells. In the zinc–carbon dry cell, the zinc shell is the anode and a carbon rod is the cathode. The newer alkaline batteries are smaller than zinc–carbon dry cells. Mercury batteries are even smaller and are used in such devices as hearing aids and calculators.

Zinc–carbon, alkaline, and mercury cells are **primary batteries,** which are not rechargeable. Batteries that are rechargeable are called **secondary batteries,** or storage batteries. A storage battery produces energy from a reversible redox reaction, which occurs in the opposite direction when the battery is recharged. A nickel–cadmium rechargeable battery is an example of a secondary battery.

▶ **Lead–acid storage batteries** The standard automobile battery is a secondary battery called a lead–acid battery. Lead is both oxidized and reduced in a lead–acid battery. A sulfuric acid solution acts as the electrolyte, so ions can migrate between electrodes. The sulfuric acid is used up as the battery generates electric current, but it is restored as the battery is recharged.

▶ **Lithium batteries** A battery that oxidizes lithium at the anode can generate a large voltage because of lithium's low E^0 value (-3.04 V). Lithium is also the lightest known metal. These properties of lithium make it desirable for use in lightweight batteries. Lithium batteries last longer than other types of batteries, and they are often used in devices such as watches, computers, and cameras.

▶ **Fuel cells** A voltaic cell that uses the oxidation of a fuel to produce electricity is called a **fuel cell.** In a hydrogen–oxygen fuel cell, hydrogen is oxidized and oxygen is reduced. The overall cell reaction is the same as the equation for the combustion of hydrogen to form water.

$$2H_2(g) + O_2(g) \rightarrow 2H_2O(l)$$

The difference is that the reaction in a fuel cell is controlled so that most of the chemical energy is converted to electrical energy instead of heat. Fuel cells produce electricity as long as fuel is supplied to them.

Practice Problems

4. What element is reduced in a standard automobile battery?

5. What term describes a battery that is not rechargeable?

6. What element is oxidized in most dry cells?

7. What is the product of the overall reaction in a hydrogen–oxygen fuel cell?

▶ **Corrosion** A redox reaction of a metal with its environment that results in the deterioration of the metal is called **corrosion.** The most common example is the rusting of iron, which occurs when air, water, and iron set up a natural voltaic cell that causes iron atoms to lose electrons. In the overall cell reaction, iron is oxidized and oxygen is reduced.

$$4Fe(s) + 3O_2(g) \rightarrow 2Fe_2O_3(s)$$

Various methods can be used to inhibit the corrosion of iron. Applying a coat of paint can help to seal out air and moisture. Iron can also be protected by placing it in contact with a piece of metal such as aluminum that is more easily oxidized than iron. The aluminum corrodes while the iron is preserved. Finally, corrosion of iron can be prevented by **galvanizing,** or coating the iron with a layer of zinc. Zinc is a self-protecting metal that oxidizes only at the surface, so the zinc coating protects the iron underneath.

21.3 Electrolysis

The use of electrical energy to bring about a nonspontaneous redox reaction is called **electrolysis.** For example, the electrolysis of water uses an electric current to decompose water into hydrogen and oxygen.

$$2H_2O(l) \xrightarrow{\text{electric current}} 2H_2(g) + O_2(g)$$

Electrolysis takes place in an **electrolytic cell.** Two of the most useful applications of electrolysis are the separation of molten sodium chloride into sodium metal and chlorine gas, and the decomposition of brine (NaCl solution) to form hydrogen gas, chlorine gas, and sodium hydroxide. Electrolysis is also used to produce aluminum metal from aluminum oxide in the Hall–Héroult process.

Chapter 21 Review

8. The following metals can be oxidized to form ions with a 2+ charge: Zn, Mg, Pt, Ca, Pb. Use Table 21-1 in your textbook to rank these metals from the most easily oxidized to the least easily oxidized.

9. Answer the following questions about this voltaic cell.

$$Cd|Cd^{2+}||Sn^{2+}|Sn$$

 a. Which electrode is the anode?

 b. Which electrode gains mass during the reaction?

 c. What is the reducing agent?

 d. What is the overall cell reaction?

 e. What is the standard cell potential?

10. The standard potential of a voltaic cell is +2.197 V, and the reduction half-reaction is $I_2(s) + 2e^- \rightarrow 2I^-(aq)$. Use Table 21-1 in your textbook to determine the probable oxidation half-reaction for this cell.

11. Name three ways to protect iron from corrosion.

12. Distinguish between a voltaic cell and an electrolytic cell.

Hydrocarbons

22.1 Alkanes

Hydrocarbons that contain only single bonds between carbon atoms are called **alkanes.** The three smallest members of the alkane series are methane, ethane, and propane. Their structural formulas are shown below. The lines indicate single covalent bonds.

$$H-\underset{\underset{H}{|}}{\overset{\overset{H}{|}}{C}}-H \qquad H-\underset{\underset{H}{|}}{\overset{\overset{H}{|}}{C}}-\underset{\underset{H}{|}}{\overset{\overset{H}{|}}{C}}-H \qquad H-\underset{\underset{H}{|}}{\overset{\overset{H}{|}}{C}}-\underset{\underset{H}{|}}{\overset{\overset{H}{|}}{C}}-\underset{\underset{H}{|}}{\overset{\overset{H}{|}}{C}}-H$$

Methane (CH_4)   **Ethane (C_2H_6)**   **Propane (C_3H_8)**

The molecular formulas of the alkanes are represented as C_nH_{2n+2}, where n is the number of carbon atoms. The names, molecular formulas, and condensed structural formulas of the first ten alkanes are listed in Table 22-1 on page 700 of your textbook.

▶ **Branched-chain alkanes** The alkanes in Table 22-1 are called straight-chain alkanes because all of their carbon atoms are bonded to each other in a single line. However, consider the compound with the following condensed structural formula.

$$\underset{CH_3CH_2\overset{\overset{\displaystyle CH_3}{|}}{CH}-\overset{\overset{\displaystyle CH_3}{|}}{CH}CH_2CH_2CH_3}{}$$

Note that the carbon atoms do not form a single line. This compound is an example of a branched-chain alkane. You can confirm that its molecular formula is C_9H_{20}, the same as for nonane. However, the compound's structure and properties are different from those of nonane. How would you determine the name of this compound? Chemists follow the rules below for naming branched-chain alkanes.

1. Determine the name of the longest continuous chain of carbon atoms. This is called the **parent chain.**
2. Number the carbon atoms in the parent chain so as to give the

lowest number to the first branch point in the chain. If the first branch points have the same number from either end of the parent chain, assign the lower number to the first branch where there is a difference in the numbering from the two ends.

3. Name the hydrocarbon groups attached to the parent chain (called **substituent groups**) by substituting *-yl* for *-ane*. For example, the group —CH_3 is called *methyl*, and —CH_2CH_3 is called *ethyl*.

4. If the same substituent group occurs more than once, use a prefix (*di-*, *tri-*, *tetra-*, etc.) before its name to indicate how many times it appears. Use numbers to show where each group is attached to the parent chain.

5. When different groups are attached to the parent chain, name the groups in alphabetical order (ignoring any prefixes).

Now recall the branched-chain alkane shown previously.

$$\begin{array}{cc} CH_3 & CH_3 \\ | & | \\ CH_3CH_2CH&-CHCH_2CH_2CH_3 \end{array}$$

The longest continuous chain contains seven carbon atoms, so the name of the parent chain is *heptane*. The parent chain can be numbered in two ways.

$$\begin{array}{cc} CH_3 & CH_3 \\ | & | \\ CH_3CH_2CH&-CHCH_2CH_2CH_3 \\ 1\ \ 2\ \ 3 & 4\ \ 5\ \ 6\ \ 7 \end{array} \qquad \begin{array}{cc} CH_3 & CH_3 \\ | & | \\ CH_3CH_2CH&-CHCH_2CH_2CH_3 \\ 7\ \ 6\ \ 5 & 4\ \ 3\ \ 2\ \ 1 \end{array}$$

The numbering on the left results in the lower number (3) for the first branch point in the chain. Both substituent groups are methyl groups, —CH_3. The name of the compound is 3,4-dimethylheptane.

Note that a comma separates the numbers, and a hyphen separates the numbers from the word. There is no space between the substituent name and the parent name. The following example problem further demonstrates how to name branched-chain alkanes.

Example Problem 22-1

Naming Branched-Chain Alkanes

Name the following alkane.

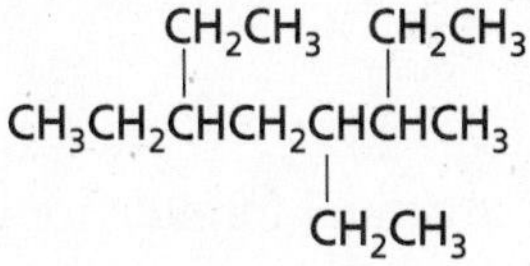

$$CH_3CH_2CHCH_2CHCHCH_3$$

with branches CH_2CH_3, CH_2CH_3 (top) and CH_2CH_3 (bottom)

a. The longest chain has eight carbon atoms, so the parent name is *octane*.

b. Number the chain in both directions.

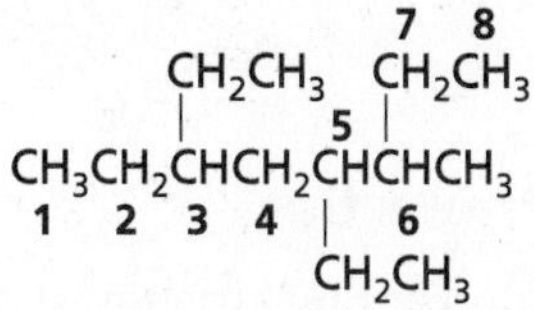

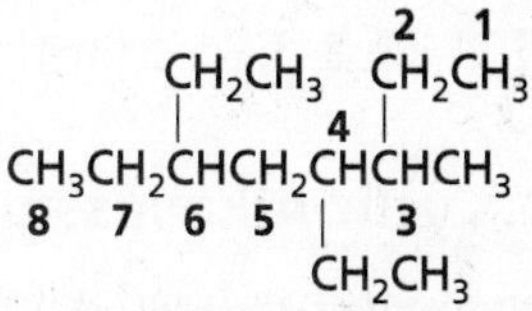

(left structure)
$$\overset{7}{CH_2CH_3} \quad \overset{8}{CH_2CH_3}$$
$$\underset{1 \; 2 \; 3 \; 4 \;\;\; 6}{CH_3CH_2CHCH_2CHCHCH_3} \;\; \overset{5}{|}$$
$$CH_2CH_3$$

(right structure)
$$\overset{2}{CH_2CH_3} \quad \overset{1}{CH_2CH_3}$$
$$\underset{8 \; 7 \; 6 \; 5 \;\;\; 3}{CH_3CH_2CHCH_2CHCHCH_3} \;\; \overset{4}{|}$$
$$CH_2CH_3$$

In both cases, the first branch point occurs at position 3. The numbering on the right results in a lower position (4) for the second branch point, so this numbering will be used.

c. Identify the attached groups. There is a methyl ($-CH_3$) group at position 3 and ethyl ($-CH_2CH_3$) groups at positions 4 and 6. The name of the alkane is 4,6-diethyl-3-methyloctane. The prefix *di-* indicates that there are two ethyl groups, and alphabetical order places *ethyl* before *methyl* in the name of the compound.

Practice Problems

1. Name the compound represented by each of the following condensed structural formulas.

a.
$$CH_3CH_2CHCH_3$$
with branch CH_3

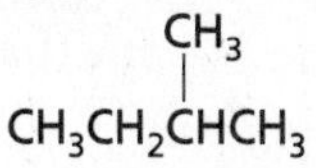

b.
$$CH_3CHCH_2CHCHCH_3$$
with branches CH_3, CH_3, and CH_2CH_3

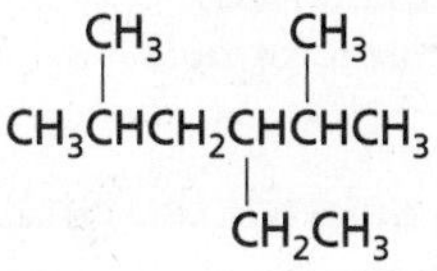

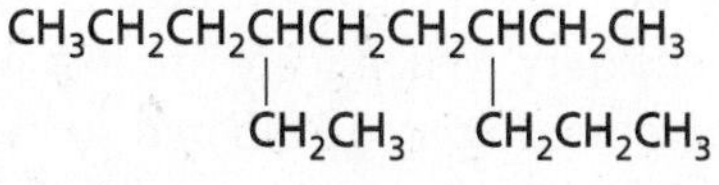

c. $CH_3CH_2CH_2CHCH_2CH_2CHCH_2CH_3$
with branches CH_2CH_3 and $CH_2CH_2CH_3$

d.

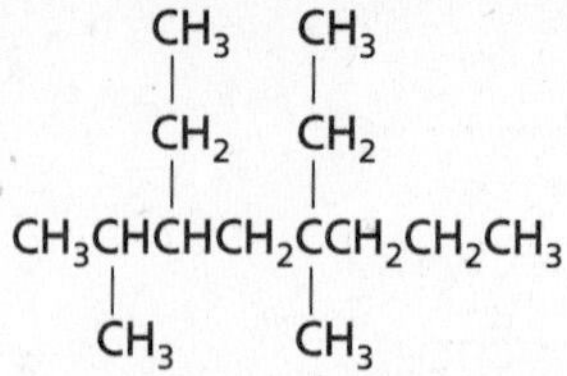

$$
\begin{array}{cc}
CH_3 & CH_3 \\
| & | \\
CH_2 & CH_2 \\
| & | \\
CH_3CHCHCH_2CCH_2CH_2CH_3 \\
| & | \\
CH_3 & CH_3
\end{array}
$$

2. Write a condensed structural formula for each of these branched-chain alkanes.

a. 3-ethylpentane

b. 2,2-dimethyl-6-propylnonane

c. 2,3,3,5-tetramethylheptane

d. 4-ethyl-3,4,5-trimethyloctane

22.2 Cyclic Alkanes and Alkane Properties

Carbon atoms can bond with each other to form ring structures. A hydrocarbon ring contains three or more carbon atoms. An alkane that contains a hydrocarbon ring is called a **cycloalkane.** Cycloalkanes have the general molecular formula C_nH_{2n}.

The name of a cycloalkane reveals the number of carbon atoms in the ring. For example, cyclopropane (C_3H_6) is the smallest cycloalkane, with three carbon atoms in the ring. Cyclobutane (C_4H_8) has a ring with four carbon atoms. A cyclic hydrocarbon can be represented by condensed, skeletal, and line structures. These structures are illustrated below for cyclobutane. Chemists usually use line structures to depict cycloalkanes.

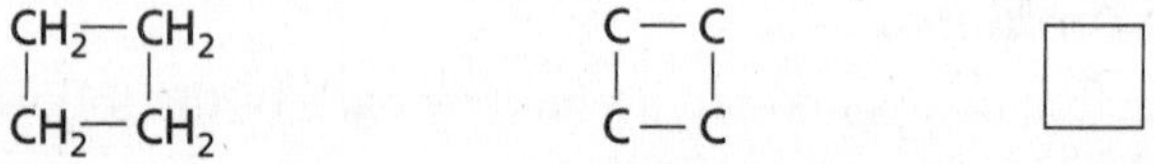

$$
\begin{array}{cc}
CH_2 - CH_2 \\
| \quad\quad | \\
CH_2 - CH_2
\end{array}
\qquad
\begin{array}{cc}
C - C \\
| \quad | \\
C - C
\end{array}
\qquad
\square
$$

| **Condensed structural formula** | **Skeletal structure** | **Line structure** |

▶ **Naming substituted cycloalkanes** Naming a cycloalkane with substituent groups is similar to naming a branched-chain alkane. The ring is always considered to be the parent chain. If there is only one substituent, no numbering is necessary. When there are two or more substituents, the carbon atoms in the ring are numbered so as to give

the lowest possible set of numbers for the substituents. The following example problem shows you how to name a substituted
cycloalkane.

Example Problem 22-2
Naming Substituted Cycloalkanes

Name the cycloalkane shown.

$$CH_3 \quad CH_2CH_3$$

$$CH_3CH_2 \quad CH_3CHCH_3$$

a. The ring has five carbon atoms, so the parent name is
cyclopentane.

b. There are four substituent groups. A methyl group and an ethyl
group are attached to the same carbon atom, and another ethyl
group is also bonded to the ring. You may be unfamiliar with
the fourth substituent group, which is called an *isopropyl* group.

c. Assign position number 1 to the carbon atom with two substituents. There are two ways to number the rest of the ring.

$$CH_3 \quad CH_2CH_3 \qquad\qquad CH_3 \quad CH_2CH_3$$
$$\begin{array}{cc} 2 & 5 \\ 3 & 4 \end{array} \qquad\qquad \begin{array}{cc} 5 & 2 \\ 4 & 3 \end{array}$$
$$CH_3CH_2 \quad CH_3CHCH_3 \qquad CH_3CH_2 \quad CH_3CHCH_3$$

Both numbering methods place branches at positions 1, 1, 3,
and 4. Should the ethyl group or the isopropyl group be
assigned position 3? The ethyl group is assigned position 3
because *ethyl* precedes *isopropyl* in alphabetical order. The
isopropyl group is assigned position 4. The numbering on the
left is correct. The name of the compound is 1,3-diethyl-4-
isopropyl-1-methylcyclopentane.

Practice Problems

3. Name the compound represented by each structure.

a.

CH$_3$CH$_2$ CH$_2$CH$_3$

b.

CH$_3$

CH$_3$CH$_2$ CH$_2$CH$_3$

CH$_2$CH$_3$

4. Draw the structures of the following cycloalkanes.

  a. propylcyclobutane

  b. 1-ethyl-2,4-dimethylcyclopentane

  c. 3-ethyl-5-isopropyl-1,1,2-trimethylcyclohexane

▶ **Alkane properties** Alkanes are nonpolar compounds with relatively low melting and boiling points compared to polar molecules of similar mass. Alkanes have low solubilities in water. They also have low reactivities, but they readily undergo combustion in oxygen and are commonly used as fuels. Alkanes are **saturated hydrocarbons** that contain only single bonds. In the next section, you will examine **unsaturated hydrocarbons,** which have at least one double or triple bond between carbon atoms.

22.3 Alkenes and Alkynes

An **alkene** is an unsaturated hydrocarbon that contains one or more double covalent bonds between carbon atoms. The general formula for alkenes with one double bond is C$_n$H$_{2n}$. The simplest alkenes are ethene and propene.

CH$_2$=CH$_2$ CH$_3$CH=CH$_2$

Ethene (C$_2$H$_4$)   **Propene (C$_3$H$_6$)**

▶ **Naming alkenes** An alkene is named by changing the -*ane* ending of the corresponding alkane to -*ene* and by specifying the location of the double bond. The carbon atoms in the parent chain are numbered so as to give the first carbon in the double bond the lowest number. This number is used in the name. Note the numbering in these two examples.

$$\overset{\text{1}\quad\text{2}\ \ \text{3}\ \ \text{4}\ \ \text{5}}{CH_2\!=\!CHCH_2CH_2CH_3}$$

1-pentene

$$\overset{\text{7}\quad\text{6}\ \ \text{5}\ \ \text{4}\quad\text{3}\ \ \text{2}\ \ \text{1}}{CH_3CH_2CH_2CH\!=\!CHCH_2CH_3}$$

3-heptene

When naming a cyclic alkene, position 1 is assigned to one of the carbons connected by the double bond, as illustrated here for the compound 1-ethyl-4-methylcyclohexene.

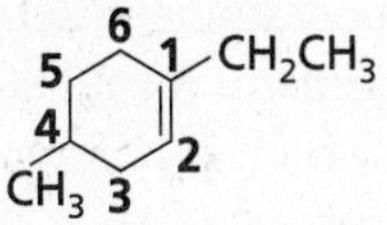

1-ethyl-4-methylcyclohexene

▶ **Naming branched-chain alkenes** The parent chain in an alkene is always the longest chain that contains the double bond, even if this is not the longest chain of carbon atoms. The numbering of the parent chain is determined by the position of the double bond. If there is more than one double bond, the positions of the double bonds are numbered so as to give the lowest set of numbers. The following example problem demonstrates how to name a branched-chain alkene.

Example Problem 22-3
Naming Branched-Chain Alkenes

Name the following alkene.

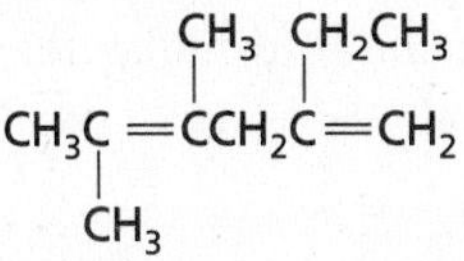

a. The longest continuous carbon chain that includes the two double bonds contains six carbon atoms. The parent name is *hexadiene*, where the prefix *di-* indicates two double bonds.

b. Numbering the chain from right to left gives the lowest numbers (1 and 4) for the double bonds.

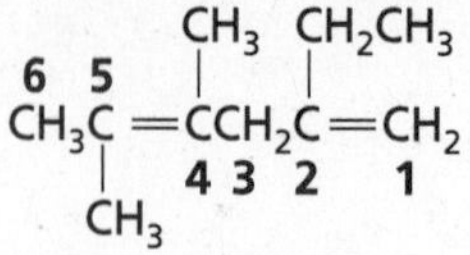

$$\underset{\underset{\displaystyle CH_3}{|}}{\overset{6\ \ 5}{CH_3C}} = \underset{4\ \ 3\ \ 2}{CCH_2}\overset{\overset{\displaystyle CH_3\ \ CH_2CH_3}{|\ \ \ \ \ \ \ |}}{C} = \overset{1}{CH_2}$$

The name of the compound is 2-ethyl-4,5-dimethyl-1,4-hexadiene.

▶ **Alkynes** An unsaturated hydrocarbon that contains one or more triple bonds between carbon atoms is called an **alkyne.** Alkynes with one triple bond have the general formula C_nH_{2n-2}. The simplest alkyne is ethyne (C_2H_2), which is commonly called acetylene. The next simplest alkyne is propyne (C_3H_4). Their condensed structural formulas are shown below.

$$CH \equiv CH \qquad\qquad CH_3C \equiv CH$$
Ethyne   **Propyne**

Alkynes are named in the same way as alkenes, with the exception that the name of the parent chain ends in *-yne* instead of *-ene*.

Practice Problems

5. Name the compounds represented by these condensed structural formulas.

a.

$$\underset{\underset{\displaystyle CH_3CH_2}{|}}{CH_3C} = \underset{\underset{\displaystyle CH_3}{|}}{CHCCH_2CH_3}$$

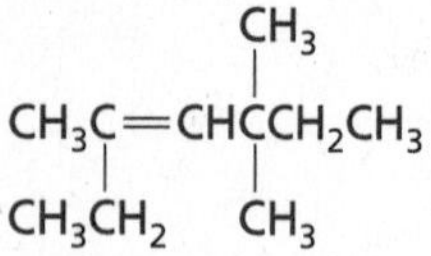

b.

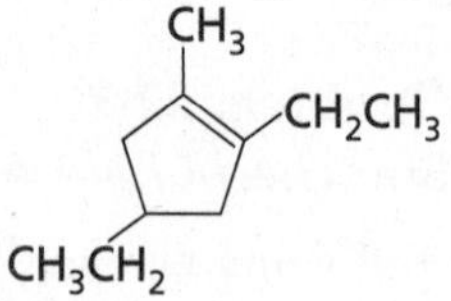

6. Draw condensed structural formulas for these compounds.

a. 3,3,4,4-tetramethyl-1-hexyne

b. 4-propyl-4-octene

c. 3-ethyl-2,6-dimethyl-2,5-heptadiene

▶ **Alkene and alkyne properties** Both alkenes and alkynes are nonpolar compounds with relatively low melting and boiling points and low solubilities in water. These properties are similar to those of alkanes. However, alkenes are more reactive than alkanes because double bonds are good sites for chemical reactivity. The triple bonds of alkynes cause them to be generally more reactive than alkenes.

22.4 Isomers

Two or more compounds that have the same molecular formula but different molecular structures are called **isomers.** The two main classes of isomers are structural isomers and stereoisomers. **Structural isomers** differ in the order in which atoms are bonded to each other. For example, hexane, 2-methylpentane, and 3-methylpentane are structural isomers because they all have the molecular formula C_6H_{14}, but they have different carbon chains. Structural isomers have different physical and chemical properties.

Isomers that have all atoms bonded in the same order but arranged differently in space are called **stereoisomers.** The two types of stereoisomers are geometric isomers and optical isomers. **Geometric isomers** result from different arrangements of groups around carbon-carbon double bonds of alkenes. The double bond prevents free rotation, so the groups attached to a double bond are locked in place. For example, examine the two geometric isomers of 2-pentene.

cis-**2-pentene**   *trans*-**2-pentene**

In the *cis*- form of 2-pentene, the substituent groups (methyl group and ethyl group) attached to the double bond are on the same side of the molecule. In the *trans*- form, they are on opposite sides of the molecule. The different arrangements of the atoms cause these two isomers to have slightly different properties.

Optical isomers are stereoisomers that result from the two possible arrangements of four different groups bonded to the same carbon atom. Optical isomers are mirror images of each other, and they have the same physical and chemical properties except in certain specialized chemical reactions.

Practice Problems

7. Decide whether the structures shown in each of these pairs represent the same compound or pairs of isomers.

a. CH$_3$CHCH$_3$ CH$_3$CH$_2$CH$_2$CH$_3$
     |
    CH$_3$

b.

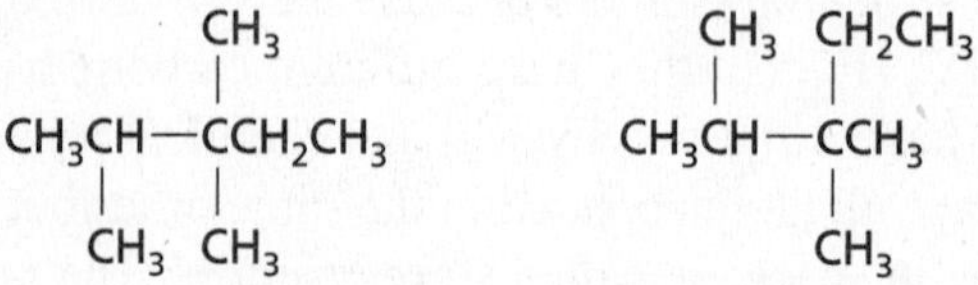

        CH$_3$ CH$_3$ CH$_2$CH$_3$
        | | |
CH$_3$CH—CCH$_2$CH$_3$ CH$_3$CH—CCH$_3$
      | | |
    CH$_3$ CH$_3$ CH$_3$

c. CH$_3$

8. Which of the following is NOT a structural isomer of 2-heptene?

a. 2,3-dimethyl-1-pentene

b. methylcyclohexane

c. 3-ethyl-2-hexene

d. 3-heptene

9. Draw the structures of *cis*-4-octene and *trans*-4-octene.

22.5 Aromatic Hydrocarbons and Petroleum

The hydrocarbon known as benzene consists of a ring of six carbon atoms with six attached hydrogen atoms. The benzene ring has a system of delocalized electrons that are shared evenly over the carbon atoms in the ring. The sharing of electrons makes benzene a fairly stable and unreactive compound. The pictorial representation of benzene usually omits the hydrogen atoms.

Benzene (C$_6$H$_6$)

Organic compounds that contain benzene rings as part of their molecular structure are called **aromatic compounds.** Nonaromatic hydrocarbons such as alkanes, alkenes, and alkynes are called **aliphatic compounds.** Some aromatic compounds contain two or more benzene rings fused together. Examples include naphthalene and anthracene. In these compounds, electrons are shared over all of the carbon atoms in the fused ring system.

Naphthalene ($C_{10}H_8$)

Anthracene ($C_{14}H_{10}$)

▶ **Substituted aromatic compounds** Aromatic compounds may have different groups attached to their carbon atoms. An example is methylbenzene, also known as toluene.

**Methylbenzene
(toluene)**

The —CH_3 group is attached to the ring in place of a hydrogen atom. You can verify that the molecular formula of toluene is C_7H_8. The rules for naming substituted benzene compounds are the same as those for naming cycloalkanes.

▶ **Sources of hydrocarbons** The major sources of hydrocarbons are petroleum and natural gas. Petroleum is a complex mixture of compounds. In a process called **fractional distillation,** petroleum is boiled and separated into components of different boiling ranges. These simpler components, called fractions, have various practical uses.

Practice Problems

10. Draw the structural formula and determine the molecular formula for these aromatic hydrocarbons.
   a. 1,2,4-trimethylbenzene
   b. 1,3-diethyl-2,5-dimethylbenzene

Chapter 22 Review

11. Why is there no compound named 2-ethylheptane? Based on this incorrect name, write the structure of the compound and provide the correct name.

12. Classify each of these compounds as saturated or unsaturated. Explain your answers.

 a. 2,3,5-trimethylhexane

 b. 1,4-pentadiene

 c. 2-heptyne

 d. isopropylcyclooctane

13. Compare the physical and chemical properties of alkanes, alkenes, and alkynes.

14. Is 2-methylcyclobutene an isomer of 1-pentyne? Explain your answer.

15. Name the compound represented by this condensed structural formula. The name should indicate whether this is the *cis-* or the *trans-* form of the compound.

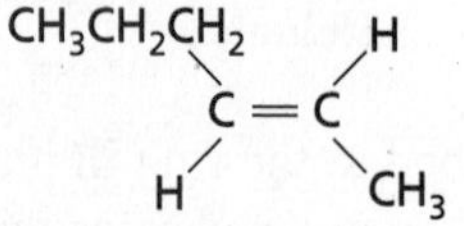

16. Which of the following are aromatic compounds? Explain your answer.

 a. 2-ethylcyclopentene

 b. 1-methyl-4-propylbenzene

 c. anthracene

 d. 3-methyl-1-hexyne

Substituted Hydrocarbons and Their Reactions

23.1 Functional Groups

A **functional group** in an organic molecule is an atom or a group of atoms that always reacts in a certain way. A **halocarbon** is any organic compound that contains a halogen substituent. An **alkyl halide** is an organic compound containing a halogen atom covalently bonded to an aliphatic carbon atom. An **aryl halide** is an organic compound containing a halogen atom bonded to a benzene ring or other aromatic group.

Halocarbons are named by the use of the prefixes *fluoro-*, *chloro-*, *bromo-*, and *iodo-* attached to the name of the alkane structure. If there is more than one kind of halogen present, the atoms are listed alphabetically. Numbers are used to identify the positions of the halogens. The chain is numbered in such a way that the lowest possible position number is given to the substituent that comes first alphabetically. In the case of aryl halides, the additional rule is that the numbering gives the substituents the lowest numbers possible.

Example Problem 23-1
Naming Halocarbons

Name the halocarbon with the following structure.

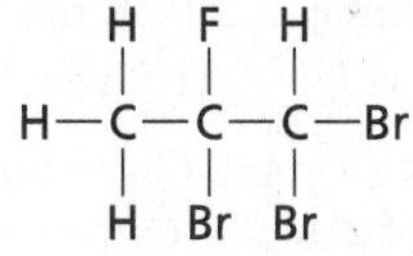

$$
\begin{array}{ccccc}
 & H & F & H & \\
 & | & | & | & \\
H- & C & -C & -C & -Br \\
 & | & | & | & \\
 & H & Br & Br &
\end{array}
$$

The alkyl chain has three carbon atoms and is therefore that of propane. The halogens present are bromine and fluorine. The prefixes to be used, in alphabetical order, are *bromo-* and *fluoro-*. The chain is numbered so that the bromine atoms, which are first alphabetically, are on the lowest-numbered carbon atoms possible. The right-hand carbon must be carbon 1. Because two of the bromine atoms are on carbon 1 and one is on carbon 2, the 1 position is indicated twice and the 2 position once. Because there are three bromine atoms, the prefix for bromine becomes *tribromo-*. The fluorine atom is on carbon 2. That number is indicated before the prefix *fluoro-*. The name of the compound is 1,1,2-tribromo-2-fluoropropane.

Practice Problems

1. Name the following halocarbons.

a.
$$\begin{array}{c} \quad\; Cl \;\; Br \;\; H \\ \quad\; | \quad\; | \quad\; | \\ H-C-C-C-I \\ \quad\; | \quad\; | \quad\; | \\ \quad\; Cl \;\; H \;\; H \end{array}$$

b.

▶ **Substitution reactions** A **substitution reaction** is one in which an atom or a group of atoms in a molecule is replaced by another atom or group of atoms. If hydrogen atoms in alkanes are replaced by halogen atoms, the substitution process is called **halogenation.** In turn, the halogen atoms in alkyl halides can be replaced by another atom or group of atoms by means of another substitution reaction.

Example Problem 23-2
Writing Equations for Substitution Reactions

Write a balanced chemical equation for the substitution reaction in which one of the hydrogens on an end carbon of butane is replaced by a chlorine atom through reaction with elemental chlorine. Name the resulting compound.

Butane is a four-carbon alkane and thus has the formula C_4H_{10}, which can be represented as $CH_3CH_2CH_2CH_3$.

Elemental chlorine has the formula Cl_2. Thus, the reactant side of the equation is $CH_3CH_2CH_2CH_3 + Cl_2$.

One of the two chlorine atoms replaces one hydrogen atom on an end carbon. The other chlorine atom combines with the displaced hydrogen atom, forming hydrogen chloride, a product. The alkyl halide formed has one chlorine atom, which is located on carbon 1. The complete equation is $CH_3CH_2CH_2CH_3 + Cl_2 \rightarrow$ $CH_3CH_2CH_2CH_2Cl + HCl$ and is balanced as written.

The halocarbon formed is 1-chlorobutane.

Practice Problems

2. Write a balanced chemical equation for each of the following substitution reactions.

a. Ethane reacts with bromine to form hydrogen bromide and an alkyl halide with one bromine atom.

b. Benzene reacts with fluorine to produce fluorobenzene and a molecule of hydrogen fluoride.

23.2 Alcohols, Ethers, and Amines

An oxygen-hydrogen group covalently bonded to a carbon atom is called a **hydroxyl group.** An organic compound in which a hydroxyl group replaces a hydrogen atom of a hydrocarbon is called an **alcohol.** Alcohols are polar and can form hydrogen bonds. As a result, they have much higher boiling points than hydrocarbons of similar shape and size. Alcohols are named using the name of the alkane, with the final -*e* replaced by -*ol*.

The position of the hydroxyl group along the alkane chain is indicated by means of a number and a hyphen at the beginning of the name. If there is more than one hydroxyl group, prefixes such as *di-* and *tri-* are placed just before the -*ol* in the name, and the -*e* is not dropped from the alkane name.

Example Problem 23-3
Naming Alcohols

Name the alcohol with the following structure.

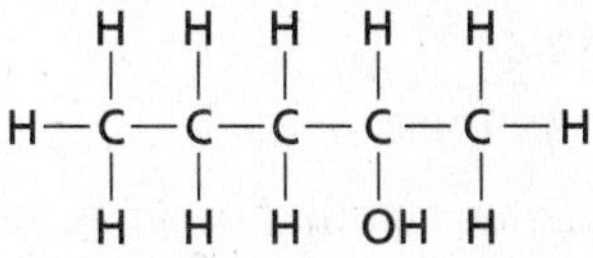

$$H-\underset{\underset{H}{|}}{\overset{\overset{H}{|}}{C}}-\underset{\underset{H}{|}}{\overset{\overset{H}{|}}{C}}-\underset{\underset{H}{|}}{\overset{\overset{H}{|}}{C}}-\underset{\underset{OH}{|}}{\overset{\overset{H}{|}}{C}}-\underset{\underset{H}{|}}{\overset{\overset{H}{|}}{C}}-H$$

The alkane chain has five carbon atoms and is therefore a pentane chain. Drop the final -*e* and add -*ol*. Counting from the right end, so as to give the hydroxyl position the lowest possible number, the hydroxyl group is on the second carbon. Thus, the prefix *2-* is used. The name of the alcohol is 2-pentanol.

Practice Problems

3. Name the alcohols with the following structures.

a.
$$H-\overset{\overset{\displaystyle H}{|}}{\underset{\underset{\displaystyle H}{|}}{C}}-\overset{\overset{\displaystyle H}{|}}{\underset{\underset{\displaystyle H}{|}}{C}}-\overset{\overset{\displaystyle H}{|}}{\underset{\underset{\displaystyle H}{|}}{C}}-\overset{\overset{\displaystyle H}{|}}{\underset{\underset{\displaystyle OH}{|}}{C}}-\overset{\overset{\displaystyle H}{|}}{\underset{\underset{\displaystyle OH}{|}}{C}}-\overset{\overset{\displaystyle H}{|}}{\underset{\underset{\displaystyle H}{|}}{C}}-H$$

b.
$$H-\overset{\overset{\displaystyle H H}{|\ \,}}{\underset{\underset{\displaystyle H}{|}}{C}}-\overset{\overset{\displaystyle H-C-H}{|\quad|}}{\underset{\underset{\displaystyle OH}{|}}{C}}-\overset{\overset{\displaystyle H H}{\ \,|}}{\underset{\underset{\displaystyle H}{|}}{C}}-H$$

4. Draw structures for the alcohols that have the following names.

  a. 2-butanol   **b.** 1,2-butanediol   **c.** 2-methyl-1-butanol

▶ **Ethers and amines** An **ether** is an organic compound that contains an oxygen atom bonded to two carbon atoms. Ether molecules cannot form hydrogen bonds with each other and thus tend to have much lower boiling points than alcohols of similar size and mass. Ethers are named by first naming the two alkyl groups that are connected to the oxygen and then adding the word *ether*. If the two groups are the same, the group is named only once, without a prefix. If the groups are different, they are named in alphabetical order.

An **amine** is an organic compound that contains nitrogen atoms bonded to carbon atoms in aliphatic chains or aromatic rings. The $-NH_2$ group in amines is called the amino group. Volatile amines tend to have offensive odors. Amines are named by first writing the name of the alkyl chain, followed by the suffix *-amine*. A number prefix indicates the position of the amino group. If there is more than one amino group, the alkane name, rather than that of the alkyl chain, is written first. Then *di-*, *tri-*, and so on are added just before the suffix *-amine*.

Example Problem 23-4
Naming Ethers and Amines

Classify each compound as an ether or an amine, and name the compound.

a.
$$H-\overset{\overset{\displaystyle H}{|}}{\underset{\underset{\displaystyle H}{|}}{C}}-O-\overset{\overset{\displaystyle H}{|}}{\underset{\underset{\displaystyle H}{|}}{C}}-\overset{\overset{\displaystyle H}{|}}{\underset{\underset{\displaystyle H}{|}}{C}}-\overset{\overset{\displaystyle H}{|}}{\underset{\underset{\displaystyle H}{|}}{C}}-\overset{\overset{\displaystyle H}{|}}{\underset{\underset{\displaystyle H}{|}}{C}}-H$$

b.
$$H-\overset{\overset{\displaystyle H}{|}}{\underset{\underset{\displaystyle H}{|}}{C}}-\overset{\overset{\displaystyle NH_2}{|}}{\underset{\underset{\displaystyle NH_2}{|}}{C}}-\overset{\overset{\displaystyle H}{|}}{\underset{\underset{\displaystyle H}{|}}{C}}-\overset{\overset{\displaystyle NH_2}{|}}{\underset{\underset{\displaystyle H}{|}}{C}}-H$$

  a. The compound is an ether because it contains an oxygen atom bonded to two carbon atoms. The left-hand alkyl group is a methyl group. The right-hand alkyl group is a butyl group. *Butyl* comes before *methyl* alphabetically; therefore, the butyl group is named first. The word *methyl* is added directly to the end of the word *butyl*. The separate word *ether* is then written. The name of the compound is butylmethyl ether.

b. The compound is an amine because it contains nitrogen in the form of amino groups. The carbon chain has four carbon atoms, making it a butyl chain. Because there is more than one amino group, the carbon part is named using the alkane name, butane. The suffix *-amine* is preceded with *tri-* because there are three amino groups. The prefixes *1*, *3*, and *3-* are added to indicate the positions of the amino groups. The name of the compound is 1,3,3-butanetriamine.

Practice Problems

5. Classify each of the following as an ether or an amine, and name the compound.

a.

$$\begin{array}{ccccc} H & H & H & H & H \\ | & | & | & | & | \\ H-C-C-C-C-C-H \\ | & | & | & | & | \\ H & H & NH_2H & NH_2 \end{array}$$

b.

$$\begin{array}{cc} H & H \\ | & | \\ H-C-O-C-H \\ | & | \\ H & H \end{array}$$

6. Write the structure for each of the following.

a. butylpropyl ether

b. 1-propylamine

c. 1,1,2,4-butanetetraamine

23.3 Carbonyl Compounds

A group made up of an oxygen atom double-bonded to a carbon atom is called a **carbonyl group.** An **aldehyde** is an organic compound in which a carbonyl group located at the end of a carbon chain is bonded to a carbon atom on one side and a hydrogen atom on the other. Aldehyde molecules tend to be polar but cannot form hydrogen bonds among themselves. Thus, aldehydes have lower boiling points than the corresponding alcohols. Aldehydes are named by replacing the final *-e* of the corresponding alkane with the suffix *-al*.

A **ketone** is an organic compound in which the carbon of the carbonyl group is bonded to two other carbon atoms. Ketones are polar and have many properties similar to those of aldehydes. They often make good solvents. Ketones are named by changing the -*e* at the end of the alkane name to -*one*. A number is used to indicate the position of the carbon atom that is double-bonded to the oxygen.

Example Problem 23-5
Naming Aldehydes and Ketones

Classify each compound as an aldehyde or a ketone, and name the following compounds.

a. 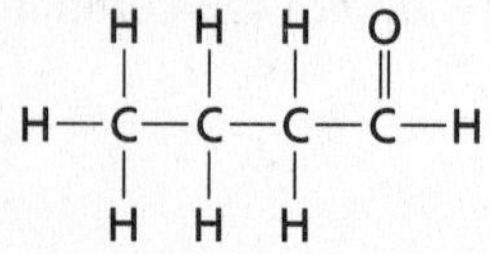 and b. 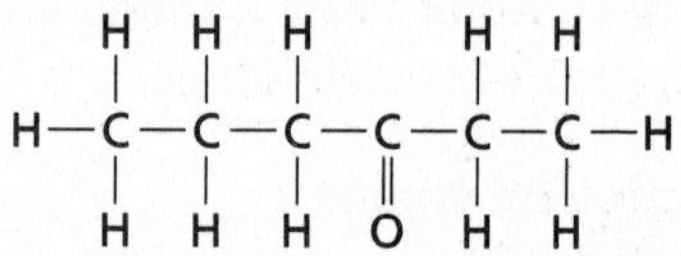 structures

a. The compound is an aldehyde because it contains a carbonyl group that is located at the end of a carbon chain and is bonded to a carbon atom on one side and a hydrogen atom on the other. Because the compound contains four carbon atoms, the corresponding alkane is butane. The suffix -*al* replaces the final -*e*. The name of the compound is butanal.

b. The compound is a ketone because the carbon of its carbonyl group is bonded to two other carbon atoms. The compound contains six carbon atoms and corresponds to the alkane hexane. The -*e* at the end of the alkane name is changed to -*one*. The carbon atom that is double-bonded to the oxygen is the third carbon from the right end; therefore, the prefix *3*- is used. The name of the compound is 3-hexanone.

Practice Problems
7. Classify each of the following as an aldehyde or a ketone, and name each compound.

a. 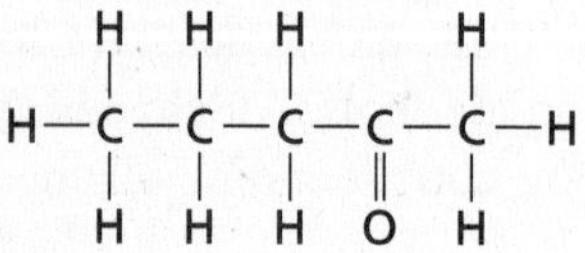 and b. 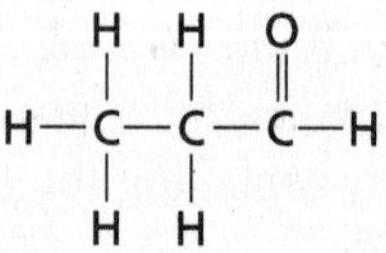 structures

8. Write the structure for each of the following.

    a. hexanal   **b.** 4-octanone

▶ **Carboxylic acids, esters, and amides** A carboxylic acid is an organic compound that has a carboxyl group. A **carboxyl group** (–COOH) consists of a carbonyl group bonded to a hydroxyl group. Carboxylic acids are polar, reactive, and acidic. They are named by changing the *-ane* suffix of the parent alkane to *-anoic acid*.

Several classes of compounds are derived by replacing the hydrogen or the hydroxyl group on the carboxyl group with a different atom or group of atoms. In one such compound, an **ester,** the hydrogen of the hydroxyl group has been replaced by an alkyl group. Esters are often polar and sweet-scented. They are named by first writing the name of the alkyl group, followed by the name of the acid, with the *-ic acid* ending replaced by *-ate*. Esters are frequently made by means of a **condensation reaction,** a reaction in which two smaller organic molecules combine to form a more complex molecule, accompanied by the loss of a small molecule such as water. In the case of esters, the reaction typically occurs between a carboxylic acid and an alcohol.

An **amide** is an organic compound in which the –OH group of a carboxylic acid is replaced by a nitrogen atom bonded to other atoms. Amide structures are found in proteins. Amides are named by writing the name of the alkane that has the same number of carbon atoms and replacing the final *-e* with the suffix *-amide*.

Example Problem 23-6
Naming Carboxylic Acids, Esters, and Amides

Classify each compound as a carboxylic acid, an ester, or an amide, and name the compound.

a. The compound is a carboxylic acid because it contains a car-boxyl group, –COOH. The compound has a total of four carbon atoms; therefore, its parent alkane is butane. The *-ane* suffix is replaced by *-anoic acid*. The name of the compound is butanoic acid.

b. The compound is an ester because the hydrogen of the hydroxyl group on a carboxylic acid has been replaced by an alkyl group. The alkyl group has two carbon atoms and is therefore an ethyl group. The acid that corresponded to the left side of the molecule has four carbon atoms and is therefore butanoic acid. The *-ic acid* ending is replaced by *-ate*. The name of the compound is ethyl butanoate.

c. The compound is an amide because it contains a carboxyl group whose hydroxyl group has been replaced by a nitrogen atom bonded to other atoms. The alkane that has the same number of carbon atoms, three, is propane. The final *-e* is replaced with the suffix *-amide*. The name of the compound is propanamide.

Practice Problems

9. Classify each of the following as a carboxylic acid, an ester, or an amide, and name each compound.

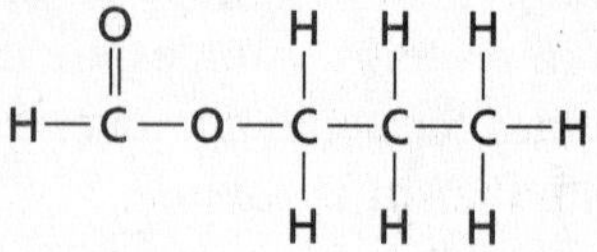

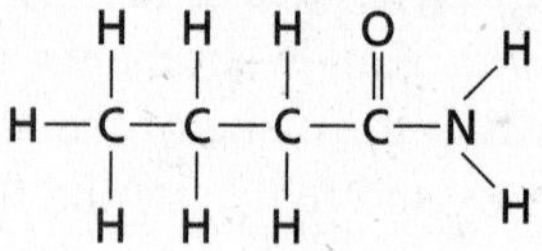

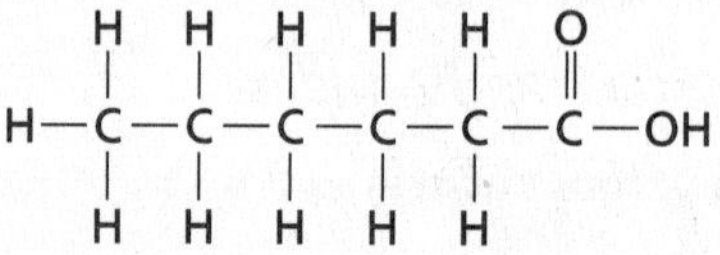

10. Explain how the ester in the preceding practice problem could be made through a condensation reaction between a carboxylic acid and an alcohol.

11. Write the structure for each of the following.

a. heptanamide

b. methyl pentanoate

c. ethanoic acid

23.4 Other Reactions of Organic Compounds

Besides substitution and condensation reactions, there also are several other important types of organic reactions. In an **elimination reaction,** a combination of atoms is removed from two adjacent carbon atoms, forming an additional bond between the carbon atoms. The formation of an alkene from an alkane is an example of an elimination reaction, in this case involving the loss of two hydrogen atoms per double bond formed. Such elimination of hydrogen atoms is called **dehydrogenation.** If the atoms removed form water, the reaction is called a **dehydration reaction.**

Another type of organic reaction is like an elimination reaction in reverse. An **addition reaction** results when other atoms bond to each of two atoms bonded by double or triple covalent bonds. A **hydration reaction** is an addition reaction in which a hydrogen atom and a hydroxyl group from a water molecule add to a double or triple bond. An addition reaction that involves the addition of hydrogen to atoms in a double or triple bond is called a **hydrogenation reaction.**

Oxidation-reduction reactions also can occur among organic molecules. Oxidation is often recognizable through a gain of oxygen or a loss of hydrogen.

Example Problem 23-7
Classifying Organic Reactions

Classify each of the following reactions as an elimination reaction or an addition reaction. Then state whether each represents hydrogenation, dehydrogenation, hydration, dehydration, or none of those.

a.
$$\underset{\underset{H}{|}}{\overset{\overset{H}{|}}{C}}=\underset{\underset{H}{|}}{\overset{\overset{H}{|}}{C} + H_2O \rightarrow H-\underset{\underset{H}{|}}{\overset{\overset{H}{|}}{C}}-\underset{\underset{H}{|}}{\overset{\overset{H}{|}}{C}}-OH$$

b.
$$H-\underset{\underset{H}{|}}{\overset{\overset{H}{|}}{C}}-\underset{\underset{H}{|}}{\overset{\overset{H}{|}}{C}}-Cl \rightarrow \underset{\underset{H}{}}{\overset{\overset{H}{}}{C}}=\underset{\underset{H}{}}{\overset{\overset{H}{}}{C}} + HCl$$

c.
$$H-\underset{\underset{H}{|}}{\overset{\overset{H}{|}}{C}}-\underset{\underset{H}{|}}{\overset{\overset{H}{|}}{C}}-\underset{\underset{H}{|}}{\overset{\overset{H}{|}}{C}}-OH \rightarrow H-\underset{\underset{H}{|}}{\overset{\overset{H}{|}}{C}}-\underset{\underset{H}{|}}{\overset{\overset{}{}}{C}}=\underset{\underset{H}{}}{\overset{\overset{H}{}}{C}} + H_2O$$

a. The reaction is an addition reaction. Atoms are bonding to each of two double-bonded carbon atoms in an ethene molecule, forming an ethanol molecule. The reaction is also a hydration reaction because a hydrogen atom and a hydroxyl group from a water molecule are being added.

b. The reaction is an elimination reaction. A combination of atoms (an H atom and a Cl atom) is being removed from two adjacent carbon atoms in a molecule of chloroethane, forming a double bond between the carbon atoms in a molecule of ethene. Because the elimination is not of two hydrogen atoms nor of water, this is neither a dehydrogenation nor a dehydration reaction.

c. The reaction is an elimination reaction. A combination of atoms (an H atom and an –OH group) is being removed from two adjacent carbon atoms in a 1-propanol molecule, forming a double bond between the carbon atoms in a propene molecule. Because the atoms removed form water, the reaction is a dehydration reaction.

Practice Problems

12. Classify each of the following reactions as an elimination reaction or an addition reaction. State further whether each represents hydrogenation, dehydrogenation, hydration, dehydration, or none of these.

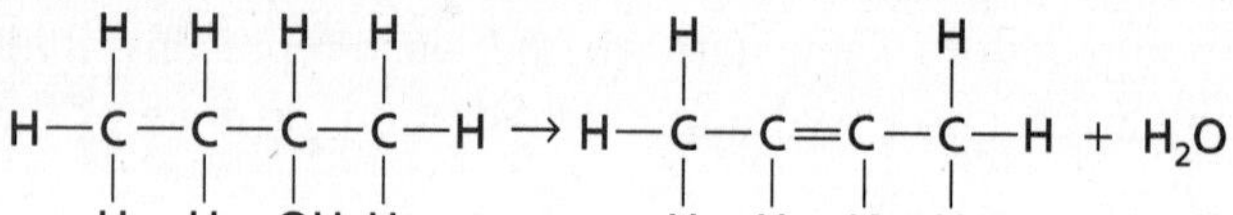

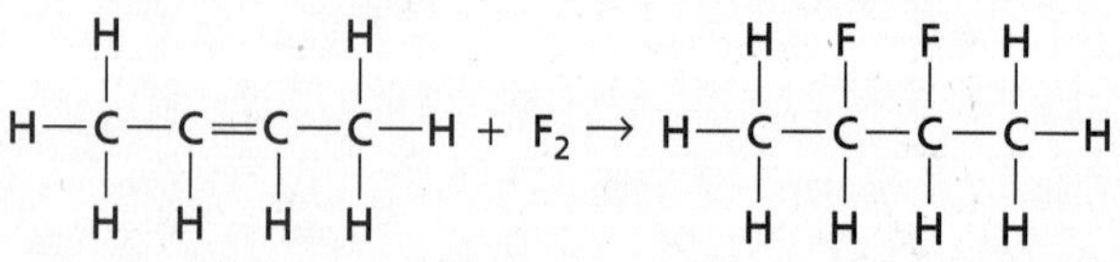

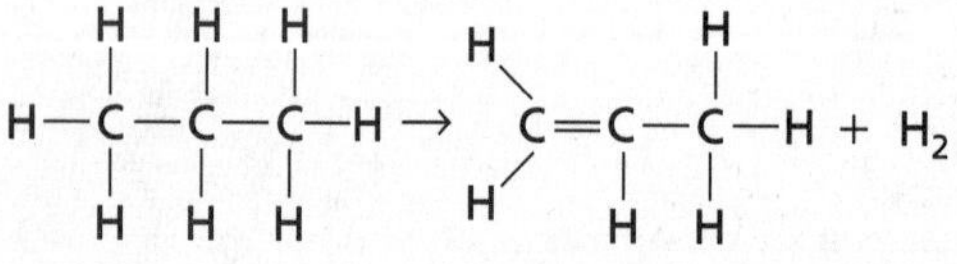

13. Predict the products of the following reactions.

   a. hydrogenation of 1-butene

   b. addition of Br_2 to a molecule of ethene

   c. dehydration of ethanol

23.5 Polymers

A **polymer** is a large molecule consisting of many repeating structural units. Polymers are made from smaller molecules called monomers. A reaction in which the monomer units are bonded together to from a polymer is called a **polymerization reaction.** The repeating group of atoms formed by the bonding of the monomers is called the structural unit of the polymer. The letter *n* is used to represent the number of structural units in a polymer chain.

In **addition polymerization,** all the atoms present in the monomers are retained in the polymer product. Polyethylene is an example of an addition polymer. Its chain contains all the atoms present in the ethene monomers that bonded to form it. **Condensation polymerization** takes place when monomers each containing at least two functional groups combine with the loss of a small by-product, usually water. Nylon is an example of a condensation polymer.

Polymers can have a wide variety of properties. Some, called **plastics,** can be heated and molded while relatively soft. A **thermoplastic** polymer is one that can be melted and molded repeatedly into shapes that are retained on cooling. A **thermosetting** polymer is one that can be molded when first prepared but when cool cannot be remelted.

Practice Problems

14. When the polymer polytetrafluoroethylene is made from the monomer tetrafluoroethene, there is no loss of atoms from each monomer molecule. Classify the polymerization as either addition or condensation.

15. When the polymer nylon-6,6 is made from the monomers adipic acid and 1,6-diaminohexane, there is a loss of water molecules. Classify the polymerization as either addition or condensation.

16. Nylon from a stocking could be remelted and molded into another form. Is nylon a thermoplastic or a thermosetting polymer?

17. Bakelite is a plastic used to make ovenproof pot handles. Is Bakelite a thermoplastic or a thermosetting polymer?

Chapter 23 Review

18. Explain how halocarbons are named, including the way in which the presence of more than one kind of halogen is indicated.

19. Contrast substitution reactions and condensation reactions, and give an example of each.

20. Compare and contrast alcohols and ethers in terms of their structures and properties.

21. Define *carbonyl group*. Explain how aldehydes and ketones differ in terms of the placement of the carbonyl group.

22. Define *carboxylic acid*. Describe the typical properties of carboxylic acids, and explain how they are named.

23. Explain how esters and amides can be made from carboxylic acids.

24. Contrast elimination reactions and addition reactions, and name two basic types of each.

25. Define *polymer*, *monomer*, and *structural unit*.

26. Contrast addition polymerization and condensation polymerization.

The Chemistry of Life

24.1 Proteins

A **protein** is an organic polymer composed of amino acids bonded together in one or more chains. An **amino acid** has a central carbon atom, to which are bonded a carboxyl group, an amino group, a hydrogen atom, and a variable side chain designated as R, as shown in the following structural formula.

Variable side chain

Amino group   Carboxyl group

Structure of an Amino Acid

Amino acids bond to each other by forming a **peptide bond,** an amide group formed by a condensation reaction between the carboxyl group of one amino acid and the amino group of another. Two amino acids linked by a peptide bond form a dipeptide. A chain of two or more amino acids linked by peptide bonds is called a **peptide.** The term **polypeptide** is applied to a chain of ten or more amino acids. Proteins may have one or several polypeptide chains, and each chain must have an exact sequence of amino acids.

Practice Problems

1. Draw the structure of a dipeptide.
2. Label the amino group and the carboxyl group of the dipeptide.
3. Draw an arrow pointing to the peptide bond.
4. Draw a square around each variable side chain.

▶ **Enzymes** Many of the proteins in an organism act as **enzymes.** These proteins catalyze chemical reactions—speeding up reactions or allowing the reactions to take place at a low temperature. The reactants in an enzyme-catalyzed process are called **substrates.** The substrate(s) bind to the enzyme at a location called the enzyme's active site, forming an enzyme-substrate complex. This interaction enables the substrate(s) to react with a much lower activation energy than they would without an enzyme.

Practice Problems

Select the term in Column B that best matches the phrase in Column A.

Column A	Column B
5. Forms when a substrate binds to an enzyme	**a.** substrate
6. Speeds up reactions	**b.** activation energy
7. A reactant in an enzyme-catalyzed process	**c.** enzyme
8. Is lowered by an enzyme	**d.** enzyme-substrate complex

24.2 Carbohydrates

Familiar **carbohydrates** include sugars, starches, and cellulose. Simple carbohydrates consist of a chain of carbon atoms having hydroxyl (–OH) groups and a carbonyl group, often in the form of an aldehyde group.

▶ **Monosaccharides** The simplest carbohydrates are the simple sugars, or **monosaccharides,** which commonly have five or six carbon atoms. Glucose, the main ingredient in corn syrup, is a familiar monosaccharide. Glucose has the molecular formula $C_6H_{12}O_6$ and can be represented by the following structures.

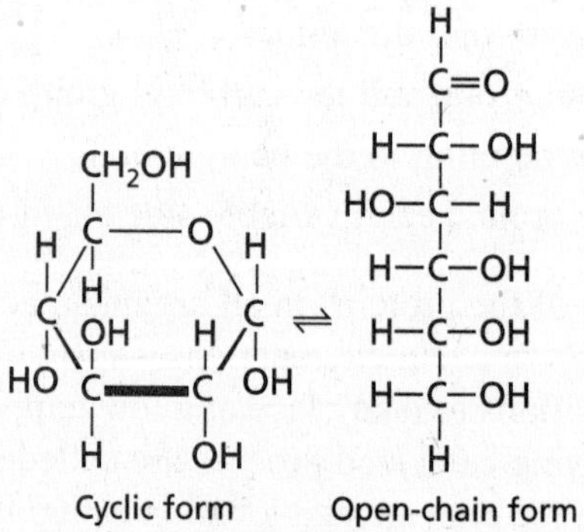

Cyclic form   Open-chain form

Glucose

Copyright © Glencoe/McGraw-Hill, a division of the McGraw-Hill Companies, Inc.

▶ **Polysaccharides** A polymer of many monosaccharides bonded into a chain is called a **polysaccharide.** Starch is a polysaccharide that consists only of glucose units. Starch is made by plants as a means of storing glucose for future use as food. Plants also link glucose units together in a different way to form the polysaccharide cellulose, which forms plant cell walls. Animals store glucose as a polysaccharide called glycogen, which is similar to starch.

Practice Problems

Select the letter of the choice that best completes the statement or answers the question.

   9. Starch is a
      a. simple sugar.   **c.** cellulose.
      b. monosaccharide.   **d.** polymer.

 10. Simple sugars do not ordinarily contain a(n) _____ group.
      a. hydroxyl   **c.** carbonyl
      b. amide   **d.** aldehyde

 11. Glucose is a(n)
      a. polysaccharide.   **c.** part of cellulose.
      b. amino acid.   **d.** 5-carbon sugar.

 12. Which of the following might be called "animal starch"?
      a. glucose   **c.** glycogen
      b. plant starch   **d.** cellulose

24.3 Lipids

Lipids are the nonpolar substances—fats, waxes, and oils—produced by living things. Lipids are not polymers, and their chemical structures vary widely.

▶ **Fatty Acids** The most familiar lipids are the plant oils and animal fats. These lipids are esters of **fatty acids,** which are carboxylic acids with long, straight hydrocarbon chains usually having between 12 and 24 carbon atoms. The simplest fatty acids are the saturated fatty acids, which have no double bonds between carbon atoms. Many other fatty acids have one or more double bonds between carbon atoms and, as a result, are unsaturated fatty acids.

▶ **Triglycerides** Animal fats and plant oils are made up primarily
of **triglycerides,** molecules in which three fatty acids are bonded
to a glycerol molecule by ester linkages, as shown in the following
diagram.

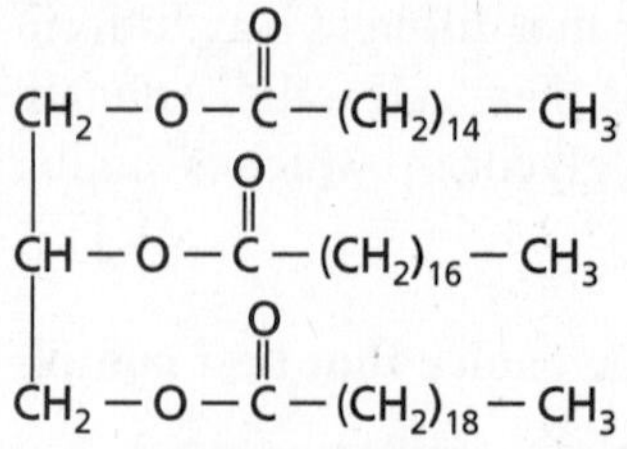

Triglyceride

Phospholipids are triglycerides in which a polar phosphate group,
instead of a third fatty acid, is bonded to the glycerol. Cell mem-
branes consist of a double layer of phospholipid molecules.

▶ **Other Lipids** Another class of lipids, **steroids,** consists of compounds whose basic structure is very different from those of other
lipids, as shown below. Cholesterol, vitamin D, and some hormones
are steroids.

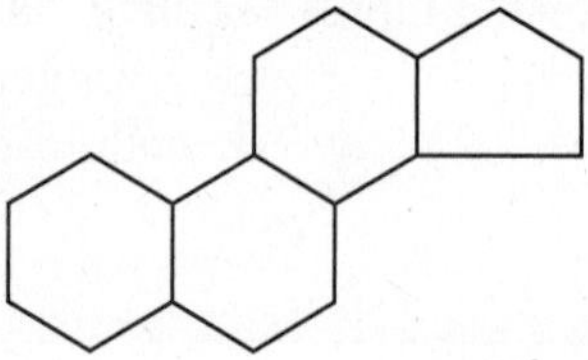

Basic Structure of a Steroid

Practice Problems

13. How do saturated and unsaturated fatty acids differ in molecular
structure?

14. How does the basic structure of steroids differ from that of fatty
acids and triglycerides?

24.4 Nucleic Acids

The sequence of amino acids in a protein is determined by the genetic information coded into long-chain polymers called **nucleic acids.** The monomers that make up nucleic acids are called **nucleotides.** Each nucleotide is made up of three parts: a phosphate group, a five-carbon sugar, and a nitrogen-containing cyclic compound called a nitrogen base. The structure of a nucleotide is shown below.

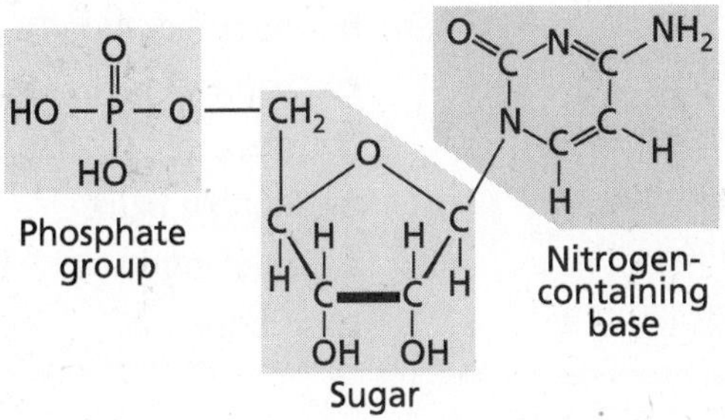

Nucleotide

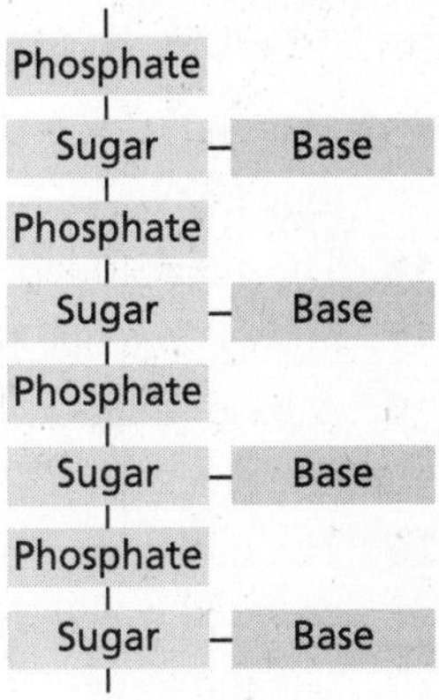

Sequence of Nucleotides

The common nucleic acids are DNA (deoxyribonucleic acid) and RNA (ribonucleic acid). These names reflect the fact that DNA contains the sugar deoxyribose and RNA contains the sugar ribose. DNA exists as a pair of polymer chains in which the backbone of each chain consists of alternating phosphate and deoxyribose units. The bases stick out from the backbone.

▶ **Base Pairing** The two chains of DNA are held together because the nitrogen bases of one chain are hydrogen-bonded to the nitrogen bases of the other chain. Because of the change in angle from one nucleotide to the next, the chains wind into a spiral called a double helix. Four different nitrogen bases are found in DNA: adenine, guanine, cytosine, and thymine. Adenine hydrogen bonds to thymine, and guanine hydrogen bonds to cytosine.

The order of these four nitrogen bases along one of the DNA chains provides the information for the sequences of amino acids in proteins. Cell mechanisms "read" the DNA sequence in groups of three bases called triplets. Each triplet codes for a specific amino acid or tells the cell to start or stop making a protein.

Practice Problems

Select the term in Column B that best matches the phrase in Column A.

Column A	Column B
15. Hold the two chains of DNA together	**a.** thymine
16. The sugar found in RNA	**b.** hydrogen bonds
17. Forms the code for one amino acid	**c.** triplet
18. A base found in DNA	**d.** protein
19. The shape of a DNA molecule	**e.** double helix
20. The molecule whose structure is encoded by DNA	**f.** nucleotides
21. The monomers in DNA and RNA	**g.** ribose

24.5 Metabolism

Metabolism refers to all the chemical reactions that take place in an organism. **Catabolism** refers to those reactions that break down large molecules—such as polysaccharides, proteins, and triglycerides—into smaller molecules. The opposite of catabolism is **anabolism,** in which smaller molecules are used to build larger molecules. In general, anabolic reactions require an input of energy, and catabolic reactions release energy.

Energy exchange in cells is carried out by molecules of **ATP** (adenosine triphosphate). To store the energy from catabolic reactions, cells carry out a reaction in which the energy is used to bond a third phosphate group to ADP (adenosine diphosphate), forming ATP. To release the stored energy for anabolic reactions, cells remove the third phosphate group from ATP, forming ADP.

▶ **Photosynthesis and Respiration** Sunlight is the source of energy for nearly all living things. During the process of **photosynthesis,** plants, algae, and some kinds of bacteria use the energy in sunlight to build carbohydrate molecules from carbon dioxide and water. Animals and some other organisms cannot photosynthesize. They take in carbohydrates by eating plants or other animals. The net chemical equation for the photosynthesis of glucose follows.

$$6CO_2 + 6H_2O + \text{light energy} \rightarrow C_6H_{12}O_6 + 6O_2$$

Carbon   Water   Glucose   Oxygen
dioxide

Later, the carbohydrates that were produced during photosynthesis are broken down in a set of catabolic reactions known as **cellular respiration,** releasing energy for life processes. The net equation for cellular respiration of glucose is the opposite of the equation for photosynthesis.

$$C_6H_{12}O_6 + 6O_2 \rightarrow 6CO_2 + 6H_2O + \text{energy}$$

Glucose   Oxygen   Carbon   Water
                             dioxide

▶ **Fermentation** When oxygen is absent or in short supply, some cells—such as yeast, some bacteria, and the muscle cells in your body—can break down glucose without oxygen in a process called **fermentation.** During fermentation, glucose is broken down and a small amount of energy is released. There are two common forms of fermentation. In alcoholic fermentation, ethanol and carbon dioxide are produced. Lactic acid fermentation produces lactic acid.

Practice Problems

Copy the following table on a sheet of paper. Then complete it by placing a check mark in the appropriate columns.

	Photosynthesis	Cellular Respiration	Fermentation
22. Releases energy			
23. Uses energy			
24. Produces glucose			
25. Breaks down glucose			
26. Produces oxygen			
27. Uses oxygen			
28. Catabolic process			
29. Anabolic process			

Chapter 24 Review

30. What effect do enzymes have on the chemical reactions that take place in living things? Why are enzymes necessary for life?

31. Corn syrup is made by breaking down cornstarch chemically. What monosaccharide would be most abundant in corn syrup? Explain your answer.

32. What four molecules would you expect to get from the breakdown of a triglyceride?

33. What is the function of DNA in the activities of a cell?

34. Your cells carry out cellular respiration. What is the function of this process? What process is the reverse of cellular respiration?

Nuclear Chemistry

25.1 Nuclear Radiation

In 1895, William Roentgen found that certain materials emit invisible rays when bombarded with electrons. He named these emissions X rays. Around the same time, Henri Becquerel was studying minerals that give off light after being exposed to sunlight—a process called phosphorescence. While trying to determine whether phosphorescent minerals also emit X rays, Becquerel found that phosphorescent uranium salts emit rays that darken photographic plates, even if they are not first exposed to sunlight. Subsequently, Marie Curie and her husband, Pierre, discovered that the darkening of the photographic plates was due to rays emitted from the uranium atoms themselves.

Marie Curie named the process by which materials emit rays radioactivity. The rays and particles that are emitted by radioactive materials are called radiation. Marie and Pierre Curie's work helped establish the field of nuclear chemistry.

▶ **Types of radiation** Isotopes of atoms with unstable nuclei are called **radioisotopes.** An unstable nucleus emits radiation to become more stable—a process called radioactive decay. Radioisotopes emit various types of radiation, including alpha particles (α), beta particles (β), and gamma (γ) rays.

Alpha radiation is a stream of alpha particles (helium nuclei). The equations below describe what happens when radium-226 (whose nucleus contains 88 protons and 138 neutrons) emits an alpha particle.

$$^{226}_{88}\text{Ra} \rightarrow {}^{222}_{86}\text{Rn} + {}^{4}_{2}\text{He}$$

Radium-226 $\rightarrow$ Radon-222 + Alpha particle

Notice that the sum of the mass numbers (superscripts) and the sum of the atomic numbers (subscripts) on each side of the arrow are equal, that is, the particles are balanced. When a radioactive nucleus undergoes alpha decay, the resulting nucleus has an atomic number that is lower by 2 and a mass number that is lower by 4. The change in atomic number changes the identity of the element.

Practice Problems

1. What element is formed when each of the following radioisotopes undergoes alpha decay? Give the atomic number and
mass number of the element.

   a. polonium-212 ($^{212}_{84}$Po)   **c.** uranium-238 ($^{238}_{92}$U)
   b. astatine-218 ($^{218}_{85}$At)   **d.** polonium-214 ($^{214}_{84}$Po)

Beta radiation is a stream of beta particles (electrons). An
example of the beta decay of iodine-131 is shown below.

$$^{131}_{53}I \rightarrow {}^{131}_{54}Xe + {}^{0}_{-1}\beta$$

Iodine-131 $\rightarrow$ Xenon-131 + Beta particle

Notice that the mass number does not change. However, the atomic
number increases by 1 (54 instead of 53), changing the identity of
the element.

The emission of a gamma ray does not change the atomic number or the mass number of a nucleus. Gamma rays almost always
accompany alpha and beta radiation. In fact, gamma rays account for
most of the energy that is lost during radioactive decay. Because
gamma rays have no effect on the atomic number or the mass number of a nucleus, it's customary to leave them out of nuclear
equations.

Practice Problems

2. What element is formed when each of the following radioisotopes undergoes beta decay? Give the atomic number and mass
number of the element.

   a. $^{3}_{1}$H   **b.** $^{235}_{92}$U   **c.** $^{60}_{26}$Fe   **d.** $^{234}_{90}$Th

25.2 Radioactive Decay

About 17 percent of known isotopes are stable. The rest decay spontaneously. The neutron-to-proton (n/p) ratio of the nucleus is a
primary factor in determining the type of radioactive decay a
radioisotope will undergo. The nucleus, which contains most of an
atom's mass, is made up of positively charge protons and neutral
neutrons. Both protons and neutrons are called **nucleons.** Protons
stay packed in the nucleus despite their strong electrostatic repulsion

forces because a **strong nuclear force,** which acts only on sub-atomic particles that are extremely close together, overcomes the electrostatic repulsion. Because neutrons have no charge but are subject to the strong nuclear force, they add an attractive force to the nucleus, which is a key to nuclear stability.

For atoms with a low atomic number (<20), nuclei with a neutron-to-proton ratio of 1:1 tend to be the most stable. Helium ($^4_2$He) is an example of an atom with a 1:1 neutron-to-proton ratio. However, as atomic number increases, more neutrons are needed to produce a nuclear force that is strong enough to balance the electrostatic repulsion forces produced by the protons. This means that the neutron-to-proton ratio increases to about 1.5:1 for the largest atoms.

Example Problem 25-1
Calculating Neutron-to-Proton Ratio

What is the neutron-to-proton ratio of $^{210}_{84}$Po?

First calculate the number of neutrons in the nucleus.

Number of neutrons = Mass number – Atomic number

Number of neutrons = 210 − 84 = 126

Then, calculate the neutron-to-proton (n/p) ratio.

$$\text{n/p ratio} = \frac{\text{Number of neutrons}}{\text{Number of protons}} = \frac{126}{84} = \frac{1.5}{1}$$

The neutron-to-proton ratio of $^{210}_{84}$Po is 1.5:1.

Practice Problems

3. Calculate the neutron-to-proton ratio for each of the following isotopes.

   a. $^{12}_{6}$C   **b.** $^{32}_{16}$S   **c.** $^{222}_{86}$Rn   **d.** $^{234}_{90}$Th

Figure 25-8 on page 811 of your textbook shows a graph of the number of neutrons versus the number of protons for all known stable nuclei. The slope of the graph shows that more neutrons are needed to stabilize a nucleus as it increases in size. The area on the graph within which all the known stable nuclei are found is called the **band of stability.** Radioactive nuclei are located outside this band. When a radioactive nucleus decays, the product is closer to or within the band of stability. All elements with an atomic number greater than 83 fall outside the band of stability and therefore are radioactive.

▶ **Types of radioactive decay** Atoms located above the band of stability usually have too many neutrons to be stable. Those located below the band of stability usually have too many protons to be stable. How does the number of neutrons and protons correlate to the type of decay a radioisotope undergoes? Beta decay decreases the number of neutrons in a nucleus and increases the stability of neutron-rich atoms by lowering the neutron-to-proton ratio. Alpha decay reduces the number of neutrons and the number of protons in a nucleus equally. Thus, the neutron-to-proton ratio does not change. Nuclei with more than 83 protons need to have both the number of neutrons and the number of protons reduced in order to become stable. These very heavy nuclei often undergo alpha decay.

Nuclei with low neutron-to-proton ratios, lying below the band of stability, are often more stable when the neutron-to-proton ratio increases. Positron emission and electron capture are two radioactive processes that increase the neutron-to-proton ratio of a nucleus.

Positron emission is a radioactive decay process that involves the emission of a positron from a nucleus. A **positron** is a particle with the same mass as an electron but with a positive charge ($_1^0\beta$). During positron emission, a proton is converted into a neutron and a positron, as shown below.

$$_1^1p \rightarrow {}_0^1n + {}_1^0\beta$$

Electron capture occurs when the nucleus of an atom draws in a surrounding electron, usually one from the lowest energy level. The captured electron then combines with a proton to form a neutron, as shown below.

$$_1^1p + {}_{-1}^0e \rightarrow {}_0^1n$$

Electron capture decreases the atomic number of the atom by one and results in the emission of an X-ray photon.

▶ **Writing and balancing nuclear reactions** Just as chemical reactions are written in balanced equations, so are nuclear reactions. Mass numbers and atomic numbers are conserved when writing a balanced nuclear reaction.

Example Problem 25-2
Balancing a Nuclear Reaction

Write a balanced nuclear equation for the alpha decay of $^{212}_{84}Po$.

Write the parts of the equation you know.

$^{212}_{84}Po \rightarrow X + ^{4}_{2}He$, where X represents the unknown product

Using the conservation of mass number yields:

Mass number of X = 212 − 4 = 208

Using the conservation of atomic number yields:

Atomic number of X = 84 − 2 = 82

Use the periodic table to identify the unknown product. $^{208}_{82}Pb$

Write the balanced equation. $^{212}_{84}Po \rightarrow ^{208}_{82}Pb + ^{4}_{2}He$

Practice Problems

4. Write a balanced nuclear equation for the alpha decay of each of the following radioisotopes.

   a. polonium-210 ($^{210}_{84}Po$)   **c.** radon-222 ($^{222}_{86}Rn$)

   b. uranium-234 ($^{234}_{92}U$)   **d.** thorium-230 ($^{230}_{90}Th$)

5. Write a balanced nuclear equation for the beta decay of each of the following radioisotopes.

   a. $^{3}_{1}H$ **b.** $^{235}_{92}U$ **c.** $^{60}_{26}Fe$ **d.** $^{234}_{90}Th$

6. Complete each of the following nuclear equations.

   a. $^{38}_{19}K \rightarrow$ _____ $+ ^{0}_{-1}\beta$

   b. $^{142}_{61}Pm + ^{0}_{-1}e \rightarrow$ _____

25.3 Transmutation

The conversion of an atom of one element to an atom of another element is called **transmutation.** All nuclear reactions except those involving only gamma emissions are transmutation reactions. Some unstable nuclei undergo transmutation naturally. Transmutation can also be induced, or forced, by bombarding a stable nucleus with radiation. The process of striking nuclei with high-velocity charged particles is called **induced transmutation.** To induce transmutation, charged particles must move at extremely high speeds in order to overcome the electrostatic repulsion between themselves and the target nucleus. Particle accelerators, also called atom smashers, are used to accelerate the particles to the extremely high speeds required.

The elements that come after uranium in the periodic table (atomic number ≥ 93) are called **transuranium elements.** These elements have all been produced in the laboratory using induced transmutation, and they all are radioactive. Scientists continue to work to produce new transuranium elements.

Example Problem 25-3
Balancing Induced Transmutation Reaction Equations

Write a balanced nuclear equation for the induced transmutation of beryllium-9 into carbon-13 by alpha particle bombardment.

Identify each participant in the reaction. Use the periodic table as needed and write the balanced equation.

$$^9_4\text{Be} + ^4_2\text{He} \rightarrow ^{13}_6\text{C}$$

Practice Problems

7. Complete the nuclear equation for each of the following induced transmutation reactions.
   a. $^{253}_{99}\text{Es} + ^4_2\text{He} \rightarrow$ _____ $+ ^1_0\text{n}$
   b. $^9_4\text{Be} + ^4_2\text{He} \rightarrow$ _____ $+ ^1_0\text{n}$

▶ **Radioactive decay rates** Radioactive decay rates are measured in half-lives. A **half-life** is the time required for one-half of a radioisotope's nuclei to decay. Each radioisotope has a different half-life. The decay of a radioisotope is described as follows.

$$\text{Amount remaining} = (\text{Initial amount})\left(\frac{1}{2}\right)^n$$

where n = the number of half-lives that has passed

Another way to write the equation is as follows.

$$\text{Amount remaining} = (\text{Initial amount})\left(\frac{1}{2}\right)^{t/T}$$

where t = elapsed time and T = the duration of the half-life

Both t and T must be written in the same units of time.

Example Problem 25-4
Calculating Amount of Remaining Isotope

Phosphorus-32 has a half-life of 14.3 days. How much of a 10.00-mg sample of phosphorus-32 will remain after 71.5 days?

First, determine the number of half-lives that has already passed.

Number of half-lives (n) = elapsed time/half-life

$$n = \frac{71.5 \text{ days}}{14.3 \text{ days/half-life}} = 5 \text{ half-lives}$$

Then, determine the amount of phosphorus-32 remaining.

$$\text{Amount remaining} = (\text{Initial amount})\left(\frac{1}{2}\right)^n$$

$$\text{Amount remaining} = (10.00 \text{ mg})\left(\frac{1}{2}\right)^5 = 0.3125 \text{ mg}$$

Practice Problems

8. Use Table 25-5 on page 818 of your textbook to answer the following questions.

   a. How much of a 2.000×10^3-mg sample of polonium-214 will remain after 1637 microseconds?

   b. How much of a 50.0-g sample of tritium will remain after 37 years?

   c. How much of a 20.0-g sample of carbon-14 will remain after one half-life? After 17 190 years?

▶ **Radiochemical dating** The process of determining the age of an object by measuring the amount of a certain radioisotope remaining in the object is called **radiochemical dating.** Carbon dating is commonly used to date things that were once living. The age of a dead organism can be estimated by comparing the decreasing ratio of unstable carbon-14 to stable carbon-12 and carbon-13 found in the organism's remains with the constant ratio found in the atmosphere. The half-life of carbon-14 is 5730 years. Thus, if an object's ratio is one-fourth that of the atmosphere, then the object is two half-lives, or 11 460 years, old. Carbon dating can be used to date only objects that are less than 24 000 years old. Other radioisotopes are used to date older objects.

25.4 Fission and Fusion of Atomic Nuclei

Albert Einstein's most famous equation relates mass and energy.

$$\Delta E = \Delta mc^2$$

In the equation, ΔE is change in energy (joules), Δm is change in mass (kg), and c is the speed of light (3.00×10^8 m/s). The equation indicates that a loss or gain in mass accompanies all chemical and nuclear reactions that produce or consume energy. Although the energy and mass changes in chemical reactions are negligible, those that accompany nuclear reactions are significant.

The energy released when an atom's nucleons bind together is called binding energy. The greater the binding energy, the more stable a nucleus is. Elements with a mass number near 60 are the most stable.

The mass of a nucleus is actually less than the sum of the masses of the nucleons. This difference is called the **mass defect.** The missing mass in the nucleus provides the energy that holds a nucleus together.

▶ **Nuclear fission** Heavy atoms (mass number > 60) tend to break into smaller atoms, thereby increasing their stability. The splitting of a nucleus into fragments is called **nuclear fission.** Nuclear fission releases a large amount of energy.

One fission reaction can lead to more fission reactions, a process called a chain reaction. A chain reaction can occur only if the starting material has enough mass to sustain a chain reaction; this amount is called **critical mass.** With a subcritical mass, the chain reaction stops or never begins. With a supercritical mass, the chain reaction accelerates and can lead to a violent explosion.

▶ **Nuclear reactors** Nuclear power plants use the process of nuclear fission to produce heat in nuclear reactors. The heat is used to generate steam, which is then used to drive turbines that produce electricity. Fissionable uranium(IV) oxide (UO_2) is commonly used as fuel in nuclear reactors. Cadmium and boron are used to keep the fission process under control. Continual adjustments are needed to keep the reaction going and under control.

▶ **Nuclear fusion** The combining of atomic nuclei is called
nuclear fusion. For example, nuclear fusion occurs within the Sun,
where hydrogen atoms fuse to form helium atoms. Fusion reactions
can release very large amounts of energy but require extremely high
temperatures. For this reason, they are also called **thermonuclear
reactions.**

Practice Problems

9. The Sun is powered by the fusion of hydrogen atoms into
helium atoms. When the Sun has exhausted its hydrogen supply, it could fuse helium-4, forming carbon-12. Write a
balanced nuclear equation for this process.

10. Write a fusion reaction that could theoretically produce one
atom of $^{275}_{113}$Uut.

25.5 Applications and Effects of Nuclear Reactions

Geiger counters, scintillation counters, and film badges are devices
used to detect and measure radiation. Geiger counters use ionizing
radiation, which produces an electric current in the counter, to rate
the strength of the radiation on a scale. **Ionizing radiation** is radiation that is energetic enough to ionize matter upon collision.

With proper safety procedures, radiation can be useful in industry, in scientific experiments, and in medical procedures. A
radiotracer is a radioisotope that emits non-ionizing radiation and
is used to signal the presence of an element or of a specific substance. Radiotracers are used to detect diseases and to analyze
complex chemical reactions.

Any exposure to radiation can damage living cells. Gamma rays
are very dangerous because they penetrate tissues and produce
unstable and reactive molecules, which can then disrupt the normal
functioning of cells. The amount of radiation the body absorbs (a
dose) is measured in units called rads and rems. Everyone is
exposed to radiation, on average 100–300 millirems per year. A dose
exceeding 500 rem can be fatal.

Chapter 25 Review

11. What is radioactivity?

12. How do neutrons affect the stability of an atom's nucleus?

13. Relate nuclear stability to an atom's size and its neutron-to-proton ratio.

14. Look at Figure 25-8 on page 811 of your textbook. Determine where on the graph each of the following isotopes would fall: above, within, or below the band of stability.

   a. $^{14}_{6}C$

   b. $^{107}_{47}Ag$

   c. an isotope with an atomic number near 50 and an n/p ratio of 1.1:1

   d. an isotope with an atomic number near 45 and an n/p ratio of 1.5:1

15. Which radioactive decay processes increase the neutron-to-proton ratio of a nucleus? Which process decreases the neutron-to-proton ratio?

16. Can carbon dating be used to date accurately the remains of an animal thought to be 1 million years old? Explain your answer.

17. What is induced transmutation? How does it relate to the transuranium elements?

18. List three devices that are used to detect and measure radiation.

19. What is the difference between nuclear fusion and nuclear fission?

20. What effect does radiation have on living cells? What dose is considered safe?

Chemistry in the Environment

26.1 Earth's Atmosphere

The **atmosphere** is a protective gaseous zone that surrounds Earth and extends into space. Chemical reactions that occur in the atmosphere help maintain a balance of atmospheric gases, but human activities, such as burning fossil fuels, can change this balance.

Earth's atmosphere is divided into five layers. The **troposphere** is closest to Earth. It is where we live and where Earth's weather occurs. Temperatures generally decrease with altitude in the troposphere. The **stratosphere,** which contains the ozone layer, is directly above the troposphere. Ozone absorbs solar radiation, which is damaging to organisms living on Earth. Beyond the stratosphere lie the mesosphere, thermosphere, and exosphere, in order. There is no clear boundary between the exosphere and outer space.

About 75 percent of the mass of all atmospheric gases, mostly nitrogen and oxygen, is found in the troposphere. The troposphere also contains dust, salts, ice, and water.

Practice Problems

1. Identify each layer of Earth's atmosphere described below.

   a. contains the air we breathe

   b. contains the ozone layer

   c. the last layer you would pass through on your way to the Moon

   d. rains here

▶ **Chemistry in the outer atmosphere** Two processes that occur in the thermosphere and exosphere—photodissociation and photoionization—shield living things on Earth from the damaging radiation that constantly bombards Earth. During photodissociation, high-energy ultraviolet (UV) solar radiation is absorbed by molecules, causing their chemical bonds to break. Molecular oxygen undergoes photodissociation, forming atomic oxygen, as shown below.

$$O_2(g) + \text{high-energy UV} \rightarrow 2O(g)$$

Photoionization occurs when a molecule or an atom absorbs enough high-energy UV solar radiation to lose an electron. Molecular oxygen undergoes photoionization, losing an electron and forming a positively charged particle, as shown below.

$$O_2(g) + \text{high-energy UV} \rightarrow O_2{}^+ + e^-$$

Practice Problems

2. Complete the chemical equation for each process below.

   a. photodissociation of nitrogen: $N_2 + \text{high-energy UV} \rightarrow$

   b. photoionization of oxygen atoms: $O + \text{high-energy UV} \rightarrow$

   c. photoionization of nitrogen: $N_2 + \text{high-energy UV} \rightarrow$

▶ **Chemistry in the stratosphere** UV radiation with the very highest energy is absorbed in the thermosphere and exosphere during photoionization. However, some UV radiation with enough energy to cause photodissociation reaches the stratosphere. In the stratosphere, O_2 absorbs the radiation, splitting into two atoms of oxygen. The oxygen atoms collide with O_2 molecules, forming energized ozone, O_3^*. (The asterisk indicates that the molecule is energized.) The energized ozone collides with another molecule (X), transferring its extra energy to X. Ozone is photodissociated back into O and O_2, as shown below.

$$O_2(g) + \text{high-energy UV} \rightarrow 2O(g)$$

$$O(g) + O_2(g) \rightarrow O_3^*(g)$$

$$O_3^*(g) + X(g) \rightarrow O_3(g) + X^*(g)$$

$$O_3(g) + \text{high-energy UV} \rightarrow O(g) + O_2(g)$$

Levels of ozone in the stratosphere have been measurably decreasing in recent decades. Chlorofluorocarbons (CFCs), such as the Freon in refrigerators and air conditioners, are thought to be responsible for this thinning of the ozone layer. Although very stable in the troposphere, CFCs photodissociate in the stratosphere, freeing chlorine atoms, which then react with ozone molecules. These Cl atoms act as a catalyst, converting O_3 to O_2, as shown below, and depleting the ozone layer.

$$CF_2Cl_2(g) + \text{high-energy UV} \rightarrow CF_2Cl(g) + Cl(g)$$

$$Cl(g) + O_3(g) \rightarrow ClO(g) + O_2(g)$$

$$ClO(g) + O(g) \rightarrow Cl(g) + O_2(g)$$

One Cl atom can exist in the stratosphere for about two years, breaking down about 100 000 ozone molecules. Although many countries have stopped making and using CFCs, it will be a long time before the ozone layer returns to former levels.

▶ **Chemistry in the troposphere** The composition of the troposphere varies from area to area, depending upon human activities. Brown, hazy smog, also known as photochemical smog, forms when sunlight reacts with pollutants that are produced largely from burning fossil fuels.

The burning of fossil fuels in internal combustion engines causes oxygen and nitrogen to react, forming nitrogen oxides, such as NO and NO_2. The NO_2, in turn, photodissociates to form atomic oxygen, which combines with molecular oxygen to form ozone. Although ozone in the stratosphere protects us from UV radiation, ozone in the troposphere irritates the eyes and causes breathing problems. Automobile exhausts also contribute unburned hydrocarbons and carbon monoxide to photochemical smog. Catalytic converters and cleaner burning engines help reduce these air pollutants.

Sulfuric acid and nitric acid form when sulfur- or nitrogen-containing pollutants in the air combine with moisture. Most of the sulfur comes from burning coal and oil. Nitrogen oxides come from automobile exhausts. The result of these reactions is acid rain, which increases the acidity of the soil and water, harming living things. Certain processes can be used to remove sulfur from coal before or during its burning, in order to reduce sulfur emissions.

26.2 Earth's Water

All of the water found in and on Earth's surface and in the atmosphere is collectively called the **hydrosphere.** Most of this water is in the oceans. Only 0.6 percent is liquid freshwater.

The water cycle is a process in which water continuously moves through the environment by the processes of evaporation, condensation, and precipitation. Liquid water on Earth evaporates. In the air, water vapor condenses on dust particles, forming clouds. Liquid water returns to Earth in the form of precipitation. Precipitation soaks into the ground and flows into bodies of water.

Practice Problems

3. Identify each process below as evaporation, condensation, or precipitation.

   a. Dew forms on blades of grass.

   b. Seawater becomes water vapor.

   c. Rain falls.

▶ **Earth's oceans** Salinity is a measure of the mass of salts dissolved in seawater, usually measured in grams of salt per kilogram of seawater. Most salt in the ocean is dissociated into ions.

The removal of salts from seawater to make it usable for living things is called **desalination.** Distillation and reverse osmosis are two ways of removing salts from seawater. During distillation, seawater is boiled to evaporate the water, leaving the salts behind. Then the pure water vapor is collected and condensed.

Reverse osmosis is more practical commercially than distillation. Seawater is forced under pressure into cylinders containing hollow semipermeable fibers. The water passes inward through the walls of the fibers, while the salt is held back. Desalinated water then flows through the inside of the fibers and is collected.

▶ **Earth's freshwater** Freshwater is a precious natural resource found mostly underground but also in lakes, rivers, and the atmosphere. Sewage, landfills, agricultural applications, and many daily activities regularly pollute our freshwater sources. Nitrogen and phosphorus, which are found in detergents, soaps, and fertilizers, are common pollutants. They encourage bacteria and algae in water to reproduce quickly. When these organisms die, their decomposition removes the oxygen from the water, killing fish and other aquatic life.

Most of the water we use daily is purified at a water-treatment plant to remove contaminants. The treatment of water in a municipal treatment plant typically includes five steps: coarse filtration, sedimentation, sand filtration, aeration, and sterilization. Sewage treatment follows steps similar to those used in water treatment.

Practice Problems

4. Identify each step in water treatment described below as one of the following: coarse filtration, sedimentation, sand filtration, aeration, or sterilization.

   a. bacteria killed in the water

   b. large solids in the water screened out

   c. water filtered through a bed of sand

   d. particles settled out of solution

   e. water sprayed into the air

26.3 Earth's Crust

Earth consists of three layers: a dense core, a thick mantle, and a thin crust. The core is further divided into a small, solid inner core and a larger, liquid outer core. The only layer of Earth that is accessible is the crust, which can be divided into solid, liquid, and gaseous parts.

The solid part of Earth's crust and the upper mantle is called the **lithosphere,** the water found in and on Earth's surface and in the atmosphere is called the hydrosphere, and the gaseous envelope around Earth is called the atmosphere. Oxygen is the most abundant element in the lithosphere; however, the lithosphere also contains alkali, alkaline earth, and transition metals.

Practice Problems

5. Identify each part of Earth's crust as solid, liquid, or gaseous.

   a. hydrosphere

   b. atmosphere

   c. lithosphere

▶ **Metals and minerals** Most metallic elements occur as minerals. A mineral is a solid inorganic compound found in nature. Minerals have distinct crystalline structures and chemical compositions.

Many metals are found in the form of oxides, sulfides, or carbonates. Oxides tend to form from transition metals on the left side of the periodic table. These elements have lower electronegativities. The elements on the right side of the table and others with higher electronegativities tend to form sulfides. Alkaline earth metals are usually carbonates.

26.4 Cycles in the Environment

The amount of matter on Earth does not change; matter is constantly recycled. Like water, a number of elements cycle through the environment.

▶ **The carbon cycle** Only about 0.03 percent of Earth's atmosphere is carbon dioxide (CO_2), yet CO_2 is vital to life. During photosynthesis, plants, algae, and some bacteria convert CO_2 from the atmosphere into carbohydrates. In turn, plants, animals, and other living things break down carbohydrates during cellular respiration and release CO_2 as a waste product. The cycle continues as photosynthetic organisms take in the released CO_2.

Carbon dioxide in the atmosphere is in equilibrium with carbon dioxide dissolved in the seas, most of which was once in the form of calcium carbonate, the main component of shells. Over time, calcium carbonate becomes limestone. When limestone is exposed to the atmosphere, it may weather and release CO_2.

Practice Problems

6. What effects might the following have on the carbon cycle? Explain your answers.

   a. the increasing size of the human population

   b. deforestation

The **greenhouse effect** is the natural warming of Earth's surface that occurs when certain gases in the atmosphere absorb some of the solar energy that is converted to heat and reflected from Earth's surface. Without this effect, Earth would be too cold to sustain its current life. The level of CO_2 in Earth's atmosphere has been increasing over the past 300 years. Less CO_2 is removed from the atmosphere because of continued destruction of forests. Thus, some scientists predict that increases in the greenhouse effect will result in a rise in global temperatures, or **global warming.**

In fact, global temperatures have increased. Scientists do not agree on the causes or the consequences of global warming. However, most concede that the consequences, such as changes in climate, can be dangerous.

▶ **The nitrogen cycle** Nitrogen makes up 78 percent of Earth's atmosphere and is a key element in living things. Because most living things cannot use nitrogen gas, it must be converted to a form living things can use, a process called **nitrogen fixation.**

There are two main ways that nitrogen fixation occurs in nature. In the atmosphere, lightning combines N_2 and O_2 into NO, which is then oxidized to NO_2. Rain converts the NO_2 to HNO_3, which falls to Earth as aqueous nitrate ions (NO_3^-). Certain bacteria living in the soil and on the roots of some plants also fix N_2 into nitrate ions. Plants take in the nitrate ions through their roots and convert them into nitrogen-containing compounds that they and other organisms need.

Organisms excrete their unused nitrogen-containing compounds as wastes. Microorganisms in the soil convert the wastes to N_2, recycling the nitrogen back into the atmosphere.

Practice Problems

7. What is the function of nitrogen fixation in the nitrogen cycle?

Chapter 26 Review

8. Contrast how photochemical smog and acid rain form.

9. List the layers of Earth's atmosphere, beginning with the layer closest to Earth.

10. Compare and contrast photodissociation and photoionization.

11. Explain how CFCs break down ozone.

12. Define *desalination*. List two methods used to accomplish desalination.

13. What is the difference between the greenhouse effect and global warming?

14. Briefly describe the two main ways that nitrogen is fixed in nature.

15. What effect would a forest fire have on the carbon cycle? Explain.

SOLVING PROBLEMS: A CHEMISTRY HANDBOOK
Answer Key

Chapter 1

Practice Problems

1. a. qualitative data   **d.** quantitative data

  b. quantitative data   **e.** qualitative data

  c. quantitative data   **f.** quantitative data

Chapter 1 Review

2. The ozone layer protects plants and animals from overexposure to ultraviolet radiation.

3. About the same time that the decrease in the ozone layer was documented, scientists noticed that a large amount of CFCs had accumulated in the atmosphere.

4. Mass is a measurement of the amount of matter in an object. Weight is a measurement of an object's mass plus the effect of Earth's gravitational pull on the object.

5. qualitative data: colorless, dissolves in water at room temperature; quantitative data: melts at 95°C, boils at 800°C

6. a. independent variable: temperature; dependent variable: ability to dissolve in water

  b. independent variable: presence of phosphorous fertilizer; dependent variable: crop growth

  c. independent variable: distance from the factory; dependent variable: acidity

7. Both hypotheses and theories are subject to revision based on new data.

8. Accept any reasonable hypotheses. Molina and Rowland hypothesized that CFCs break down in the atmosphere due to the Sun's ultraviolet radiation and that a chlorine particle produced by the breakdown of CFCs could break down ozone.

9. a. applied   **b.** pure   **c.** applied

10. Answers should include any two of the rules listed in the Safety in the Laboratory table that relate to the handling of chemicals (rules 4, 10–18, 22, and 23).

11. according to your teacher's directions

Chapter 2

Practice Problems

1. 100 centigrams

2. 1000 liters

3. 1 000 000 000 nanoseconds

4. 1000 meters

5. 1.9 g/cm^3

6. 1.6 g/mL, or 1.6 g/cm^3

7. 0.862 g/mL, or 0.862 g/cm^3

8. silver

9. 2.2 cm^3

10. 17 g

11. **a.** 327 K
    b. 219 K
    c. 288 K

12. **a.** −241°C
    b. −273°C
    c. 8°C

13. **a.** 5×10^4 m/s^2
    b. 6.2×10^{-10} kg
    c. 2.3×10^{-5} s
    d. 2.13×10^7 mL
    e. 9.909×10^8 m/s
    f. 4×10^{-9} L

14. **a.** 4.62×10^{21}
    b. 6.17×10^8
    c. 2.280×10^5
    d. 1.67×10^{-4}
    e. 5.92×10^{-5}
    f. 8.198×10^2

15. **a.** 6×10^2 m^2
    b. 9×10^{14} km^2
    c. 4×10^{-8} mm^2
    d. 1×10^3 kg/L
    e. 4 m/s
    f. 4×10^4 km/s

16. 884 700 cm

17. 1560 mm

18. 11 L

19. 168 hr; 10 080 min

20. 0.783%

21. 2.00%

22. **a.** 6.00%
    b. The measurements are extremely precise but not accurate.

23. **a.** 2
    b. 4
    c. 5
    d. 4

24. **a.** 2.4995×10^{-4}
    b. 9.0708×10^2
    c. 2.4502×10^7
    d. 3.0010×10^8

25. **a.** 439 g
    b. 695.91 mL
    c. 510 km^2
    d. 588.1 m/s

26. a. 1.5 g/cm^3 **c.** 9.8 m/s

   b. 1.10 g/cm^3 **d.** 2.8°C/kJ

Chapter 2 Review

27. a. mL or L   **b.** km   **c.** g

28. It multiplies the quantity by 1000.

29. g/cm^3 or g/mL; they are derived units because they involve a combination of base units.

30. accurate, because the target changes with each throw

31. The accepted value; for a large value, the measurements might be precise. For a small value, they would not be.

32. $\dfrac{1 \text{ hour}}{60 \text{ minutes}}$

33. a. circle or bar graph   **d.** circle graph

   b. bar or line graph   **e.** line graph

   c. bar graph   **f.** line graph

Chapter 3

Practice Problems

1. a. chemical   **2. a.** liquid, gas   **3. a.** physical

   b. physical   **b.** gas   **b.** chemical

   c. physical   **c.** solid   **c.** chemical

   d. chemical   **d.** solid, liquid   **d.** physical

   e. physical   **e.** gas   **e.** physical

   f. chemical   **f.** liquid   **f.** chemical

   g. physical

   h. chemical

4. The tube will have a mass of 18.48 g, the same mass it had before the reaction. The tube was sealed, and no mass is lost or gained in a chemical reaction.

5. 4.73 g

6. 12.05 g barium nitrate

7. 33.38 g sodium chloride

8. a. homogeneous   **d.** homogeneous
   b. heterogeneous   **e.** homogeneous
   c. heterogeneous   **f.** heterogeneous

9. a. compound   **e.** element
   b. element   **f.** element
   c. compound   **g.** compound
   d. element

10. 72.9% bromine; 100%

11. 4.48% hydrogen, 60.00% carbon, 35.52% oxygen

12. 33.6% sulfur, 66.4% copper

13. 14.6 g oxygen

Chapter 3 Review

14. The composition of a substance is always the same. The composition of a mixture such as lemonade can vary greatly.

15. a. physical   **c.** physical
    b. chemical   **d.** chemical

16. Yes; the physical properties, such as shape, texture, and physical state, of a pure substance can vary. No; a substance has a fixed composition.

17. Solid: particles close together and in a regular arrangement; Liquid: particles close together and free to pass one another; Gas: particles far apart and free to move in all directions

18. The mass of the nail will be greater than 13.8 g because the iron has combined chemically with oxygen from the air. The added oxygen increases the mass of the nail.

19. a. chemical   **e.** physical
    b. physical   **f.** physical
    c. chemical   **g.** physical
    d. chemical

20. Aluminum must be an element because it cannot be separated into simpler substances. All elements are substances.

21. Yes; cassiterite is a compound, which is a substance. A substance has the same composition everywhere it is found.

22. No; the properties of a compound usually differ from those of the individual elements that make up the compound. In this case, the compound carbon disulfide is a foul-smelling liquid at room temperature.

Chapter 4

Practice Problems

1. **a.** 1   **d.** 1   **g.** 1   **j.** 2
   b. 3   **e.** 3   **h.** 1   **k.** 4
   c. 4   **f.** 4   **i.** 4   **l.** 3

2. **a.** $33\ p^+, 33\ e^-$ **d.** $42\ p^+, 42\ e^-$
   b. $79\ p^+, 79\ e^-$ **e.** $84\ p^+, 84\ e^-$
   c. $9\ p^+, 9\ e^-$ **f.** $56\ p^+, 56\ e^-$

3. **a.** selenium   **d.** promethium
   b. boron   **e.** plutonium
   c. gallium

4. **a.** tungsten   **d.** ytterbium
   b. calcium   **e.** neptunium
   c. indium

5. argon

6. The atoms have differing numbers of neutrons. Cr-50 has 26 neutrons, Cr-52 has 28, and Cr-54 has 30.

7. $9\ p^+, 9\ e^-, 10\ n^0, {}^{19}_{9}\text{F}$

8. The nucleus of the atom is composed of $40\ p^+$ and $54\ n^0$. Around the nucleus are $40\ e^-$.

9. cadmium; $48\ p^+$ and $48\ e^-$

10. platinum; $78\ p^+, 120\ n^0$

11. 69.723 amu

12. 72.59 amu

13. No bromine-80 is found in nature. Naturally occurring bromine consists of an almost 50:50 mix of Br-79 and Br-81, resulting in an average atomic mass close to 80 amu.

14. 36.97 amu; accepted value = 36.966 amu

Chapter 4 Review

15. The ideas of the ancient Greeks were based on observation, thought, and debate. Dalton's ideas were based on experimental results.

16. Electrons were found to have negative charge. The fact that atoms have no charge was evidence that balancing positive charges are present.

17. In Rutherford's gold-foil experiment, most of the alpha particles passed straight through the gold foil as though they were passing through empty space. However, a few particles were deflected, and some bounced straight back, indicating the existence of a small, massive body surrounded by empty space. Protons and neutrons are responsible for the mass of the nucleus.

18. Argon atoms have 18 protons and 18 electrons. An argon atom is neutral, so the number of electrons must equal the number of protons.

19. No; the atoms of different elements have different numbers of protons.

20. An atom of selenium-80 has a nucleus with 34 protons and 46 neutrons, and 34 electrons surround the nucleus.

21. This number is the atomic mass of the element. It is the weighted average of the masses of all of the naturally occurring isotopes of the element.

22. $^{123}_{51}$Sb, $^{28}_{14}$Si, $^{199}_{80}$Hg

23.

Types of Radiation from Radioactive Elements			
Type of radiation	Mass number	Charge	Composition
Beta	1/1840	1−	electron
Alpha	4	2+	$2\,p^+, 2\,n^0$
Gamma	0	none	energy only

Chapter 5

Practice Problems

1. 4.74×10^{14} Hz

2. 6.25×10^{-10} m; (0.625 nm)

3. 3.05 m

4. 3.33×10^{-13} J

5. $E_{photon\,(violet)} = 4.5 \times 10^{-19}$ J
   $E_{photon\,(red)} = 2.8 \times 10^{-19}$ J
   $E_{difference} = 1.7 \times 10^{-19}$ J

6. First calculate the frequency of the light.
   $\nu = 6.12 \times 10^{15}$ Hz
   $E_{photon} = 4.06 \times 10^{-18}$ J

7. Bohr's model showed that electrons move in circular orbits around the nucleus. The electrons have the energies associated with those orbits.

8. The electron is moving from an orbital of higher energy to an orbital of lower energy.

9. The second principal energy level can hold eight electrons (two in the 2s sublevel and six more in the 2p sublevel). The third principal energy level can hold eighteen electrons (two in the 3s sublevel, six in the 3p sublevel, and ten in the 3d sublevel). The difference is due to the fact that the larger third principal energy level has one more available energy sublevel.

10. The sublevels are written in order of increasing energy.

11. The number of electrons is indicated as a superscript integer to the right of the sublevel.

12. a. $1s^2 2s^2 2p^6 3s^2 3p^4$

c. $1s^2 2s^2 2p^6 3s^2 3p^6 4s^2 3d^{10} 4p^5$

b. $1s^2 2s^2 2p^6 3s^2 3p^6 4s^2$

d. $1s^2 2s^2 2p^6 3s^2$

13. a. $[He]2s^2 2p^5$

f. $[Kr]5s^2 4d^5$

b. $[Ne]3s^2 3p^3$

g. $[Kr]5s^2 4d^{10} 5p^5$

c. $[Ar]4s^2$

h. $[Xe]6s^2 4f^{11}$

d. $[Ar]4s^2 3d^7$

i. $[Xe]6s^2 4f^{14} 5d^7$

e. $[Ar]4s^2 3d^{10} 4p^4$

j. $[Rn]7s^2$

14. a. $\cdot \ddot{N}\colon$ **c.** $\colon\!\ddot{N}\!e\colon$ **e.** $\cdot \ddot{S}b\colon$ **g.** $\cdot Pb\cdot$

b. $\cdot \dot{A}l\cdot$ **d.** $\cdot Sr\cdot$ **f.** $\colon\!\ddot{\ddot{I}}\cdot$ **h.** $Cs\cdot$

15. eight dots surrounding the symbol for the element

16. Student responses may include any of the group 5A elements: nitrogen, phosphorus, arsenic, antimony, and bismuth.

Chapter 5 Review

17. The equation $c = \lambda \nu$ relates the speed, wavelength, and frequency of an electromagnetic wave. The frequency of a given wave cannot change; wave speed and wavelength are directly related. Therefore, the wavelength will decrease as speed decreases.

18. Light striking a metal plate dislodges electrons from the plate. Einstein concluded that light could act not only as a wave, but also as a stream of particles, called photons.

19. Energy increases as frequency increases. Energy decreases as wavelength increases (because frequency decreases).

20. According to Bohr's model, an atom's electrons travel around the nucleus in circular orbits, each of which has a specific energy value. Electrons can jump only from one allowable orbit to another. Therefore, electrons can increase or decrease in energy only by certain amounts. These amounts correspond to the frequencies (colors) of the light seen in the atomic emission spectrum.

21. Three energy sublevels are available—s, p, and d. An s sublevel can hold two electrons; a p sublevel can hold six electrons; and a d sublevel can hold ten electrons.

22. In the modern model of the atom, electrons are not in specific orbits. Instead, they are in regions of space in which there is a high probability of finding electrons of a certain energy.

23. a. fluorine   **e.** rhodium
   b. manganese   **f.** polonium
   c. germanium   **g.** sulfur
   d. sodium   **h.** cadmium

24. The dots represent the electrons in the highest principal energy level of the atom. These electrons are called valence electrons, and they determine many of the chemical properties of an element.

25. Student responses may include helium and any of the group 2A elements: beryllium, magnesium, calcium, strontium, barium, and radium.

Chapter 6

Practice Problems

1. c, f   **7.** d, f   **9. a.** 3
2. g   **8. a.** 2   **b.** 4
3. a, f   **b.** 6   **c.** 5
4. f   **c.** 4   **d.** 4
5. b, f   **d.** 5   **e.** 5
6. e, f   **e.** 7

10. a. tin   **d.** arsenic
   b. bismuth   **e.** germanium
   c. helium

11. **a.** group 7A, second period, p block

b. group 2A, fourth period, s block

c. group 5A, fifth period, p block

d. group 5B, fourth period, d block

12. **a.** $[Xe]6s^2$ **d.** $[Ar]4s^23d^3$

b. $[Ne]3s^23p^5$ **e.** $[Kr]5s^1$

c. $[Ne]3s^23p^2$

13. **a.** Sr   **d.** Ge

b. Sr   **e.** W

c. Sn

14. The energy should increase because, as atoms become smaller across a period, the electrons being removed are closer to the positive nucleus.

15. **a.** Mg   **d.** I^-

b. S^{2-} **e.** Na^+

c. Ba^{2+}

16. The S^{2-} ion is larger. Both Mg and S are in the same period, but S^{2-} has gained electrons, attaining an argon configuration, whereas Mg^{2+} has lost two electrons, attaining a neon configuration. Therefore, S^{2-} is larger.

17. **a.** Mg   **c.** Ca   **e.** Al

b. O   **d.** Cl   **f.** Br

18. **a.** Ca   **c.** Na   **e.** Sr

b. I   **d.** K   **f.** As

19. **a.** lose 1   **d.** lose 2   **g.** Argon does not lose or gain electrons because it is a noble gas with a stable octet configuration.

b. gain 1   **e.** lose 3

c. gain 2   **f.** gain 1

20. **a.** Ar   **c.** Ar   **e.** Rn   **g.** Xe

b. Kr   **d.** He   **f.** Ne

21. **a.** Mg   **c.** Cl   **e.** O

b. Al   **d.** Ca   **f.** Br

Copyright © Glencoe/McGraw-Hill, a division of the McGraw-Hill Companies, Inc.

Chapter 6 Review

22. The table is periodic because it shows a repeating pattern of properties with each succeeding row of elements.

23. The elements in a group have similar arrangements of valence electrons.

24. The common property of these elements, the noble gases, is their lack of chemical reactivity. They are unreactive because they have a stable octet of eight valence electrons.

25. The number indicates the number of valence electrons—the number of electrons in s and p orbitals in the highest energy level.

26. **a.** Atomic radius generally decreases across a period and increases through a group of elements.

b. Electronegativity generally increases across a period and decreases through a group of elements.

c. First ionization energy generally increases across a period and decreases through a group of elements.

d. The radius of positive ions generally decreases across a period. Likewise, the radius of negative ions decreases across a period. The ionic radius increases through a group of elements.

27. The higher the electronegativity value of an element, the greater is the tendency of that element to gain electrons when forming a chemical bond.

Chapter 7

Practice Problems

1. **a.** Both calcium and strontium are in group 2A, have two valence electrons, and form 2+ ions; strontium has a lower ionization energy and greater reactivity.

b. Both lithium and francium are in group 1A, have one valence electron, and form 1+ ions; francium has a lower ionization energy and greater reactivity.

   c. Cesium is in group 1A, has one valence electron, and forms
      1+ ions; barium is in group 2A, has two valence electrons,
      and forms 2+ ions; cesium has a lower ionization energy
      and greater reactivity.

2. a. magnesium   **b.** beryllium

3. a. As: arsenic, group 5A, three valence electrons, metalloid;
       Bi: bismuth, group 5A, five valence electrons, metal

   b. Ge: germanium, group 4A, four valence electrons, metalloid;
       N: nitrogen, group 5A, five valence electrons, nonmetal

   c. B: boron, group 3A, three valence electrons, metalloid; Sn:
       tin, group 4A, four valence electrons, metal

4. group 3A, thallium

5. a. Se: selenium, group 6A, six valence electrons, 2− ions, used
       in solar panels and photocopiers; Cl: chlorine, group 7A,
       seven valence electrons, 1− ions, used as a bleach and disin-
       fectant and to make certain plastics

   b. I: iodine, group 7A, seven valence electrons, 1− ions, silver
       compound used to coat photographic film; He: helium,
       group 8A, two valence electrons, no ions typically formed,
       used in balloons and by divers

   c. S: sulfur, group 6A, six valence electrons, 2− ions, used to
       make sulfuric acid; F: fluorine, group 7A, seven valence
       electrons, 1− ions, used in toothpaste to protect tooth
       enamel

6. group 8A, noble gases, argon

7. group 6A, polonium

8. Nickel: period 4, d block, eight d electrons, two of them
    unpaired, ferromagnetic; yttrium: period 5, d block, one
    d electron, unpaired, paramagnetic

9. a. period 6, f block, lanthanide
   b. period 7, f block, actinide
   c. period 6, d block, transition metal

Chapter 7 Review

10. Elements within the same group of the periodic table have the same number of valence electrons, but different numbers of nonvalence electrons.

11. A diagonal relationship is a similarity in properties between certain period 2 elements and an element in period 3 of an adjacent group in the periodic table. An example is the relationship between lithium and magnesium.

12. The alkali metals have one valence electron and are in group 1A. The alkaline earth metals have two valence electrons and are in group 2A. Both types of metals are reactive, but the alkaline earth metals are less reactive. The alkali metals are softer.

13. Carbon is a nonmetal, silicon is a metalloid, and lead is a metal. All have four valence electrons and are located in group 4A.

14. Allotropes are forms of an element in the same state that have different structures and properties. Students' examples will vary and may include the allotropes of carbon, phosphorus, oxygen, or sulfur.

15. Nitrogen (group 5A) has five valence electrons and is relatively unreactive. Oxygen (group 6A) has six valence electrons and is reactive. Fluorine (group 7A) has seven valence electrons and is extremely reactive. Neon (group 8A) has eight valence electrons and is extremely unreactive.

16. The transition metals are elements whose final electron enters a d sublevel. They are located in the B groups of the periodic table. The inner transition metals, the lanthanides and actinides, are elements whose final electron enters an f sublevel. They are located below the main body of the periodic table. Students' examples will vary.

Chapter 8

Practice Problems

1. neutral Br: $1s^2 2s^2 2p^6 3s^2 3p^6 4s^2 3d^{10} 4p^5$; ion: Br^-, anion, $1s^2 2s^2 2p^6 3s^2 3p^6 4s^2 3d^{10} 4p^6$

2. neutral Ga: $1s^22s^22p^63s^23p^64s^23d^{10}4p^1$; ion: Ga^{3+}, cation, $1s^22s^22p^63s^23p^63d^{10}$

3. neutral S: $1s^22s^22p^63s^23p^4$; ion: S^{2-}, anion, $1s^22s^22p^63s^23p^6$

4. neutral Rb: $1s^22s^22p^63s^23p^64s^23d^{10}4p^65s^1$; ion: Rb^+, cation, $1s^22s^22p^63s^23p^64s^23d^{10}4p^6$

5. Al: $[Ne]3s^2\ 3p^1$; F: $[He]2s^2\ 2p^5$; one Al for every three F

6. Li: $[He]2s^1$; O: $[He]2s^22p^4$; two Li for every O

7. Be: $[He]2s^2$; Se: $[Ar]4s^23d^{10}4p^4$; one Be for every Se

8. Ga: $[Ar]4s^23d^{10}4p^1$; S: $[Ne]3s^23p^4$; two Ga for every three S

9. a. not ionic **b.** not ionic **c.** ionic **d.** ionic

10. a. LiF   **b.** MgS   **c.** MgO

11. Na_2S

12. Mg_3N_2

13. K_3P

14. BaF_2

15. AlN

16. Na_3PO_4

17. $(NH_4)_2CO_3$

18. $Al_2(CrO_4)_3$

19. $Ca(OH)_2$

20. ammonium iodide

21. sodium bromate

22. magnesium nitrate

23. potassium hydrogen sulfate

24. ammonium perchlorate

25. aluminum hypochlorite

26. iron(II) fluoride

27. not conclusively, because ionic solids also conduct electricity when melted

28. malleability; no, because metals are malleable

29. nickel, because nickel is a metal and metals are good conductors of electricity; ductility, malleability, metallic luster, very high boiling point

30. an interstitial alloy

31. a substitutional alloy

Chapter 8 Review

32. A chemical bond is a force that holds two atoms together. Elements tend to form bonds so as to achieve the stable electron configuration of a noble gas.

33. A cation is a positively charged ion formed when one or more electrons are transferred from an atom. An anion is a negatively charged ion formed when one or more electrons are transferred to an atom. An example of a cation is the sodium ion, Na^+. An example of an anion is the chloride ion, Cl^-.

34. An ionic bond is the electrostatic force that holds oppositely charged ions together in an ionic compound. A potassium atom transfers its valence electron to an iodine atom, resulting in a potassium cation and an iodide anion, which form an ionic bond because of their opposite charges.

35. Ionic compounds are made up of a three-dimensional crystal lattice, with positive ions surrounded by negative ions, and vice versa. Typical properties include high melting and boiling points, brittleness, inability to conduct electricity when solid, and ability to conduct electricity when melted or dissolved in water.

36. Determine the charge of each type of ion, generally by referring to the periodic table or a table of common ions. Then choose numbers of positive and negative ions such that the sum of the positive and negative charges is zero. Write the symbols for the atoms and use subscripts to show the number of each ion per formula unit.

37. If there are two oxyanions, the one with fewer oxygen atoms per ion is named using the suffix *-ite*. The name of the oxyanion with more oxygen atoms has the suffix *-ate*. In the case of the halogens, if there is one oxygen atom, use the prefix *hypo-* and the suffix *-ite*. If there are two oxygens, use the suffix *-ite*. If three, use the suffix *-ate*. If four, use the prefix *per-* and the suffix *-ate*.

38. The valence electrons are delocalized, forming a mobile sea of electrons that attract the metallic cations. The mobility of the electrons accounts for malleability and ductility, which require

the particles to move past each other, and for the ability to conduct electricity, which requires electrons to flow easily. The
delocalized electrons also interact with light, which accounts
for metallic luster. Boiling points tend to be extremely high
because of the difficulty of completely removing atoms from
the attracting electrons and cations. Melting points are considerably lower because of the ability of the particles to move past
each other.

39. Both are mixtures that have metallic properties. In substitutional
alloys, atoms of the original metal have been replaced by other
metal atoms of similar size. In interstitial alloys, small holes in
a metallic crystal have been filled with smaller atoms.

Chapter 9

Practice Problems

1. a. 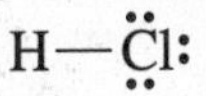 H—C̈l: one single covalent bond, three lone
pairs on Cl, no lone pairs on H

b. 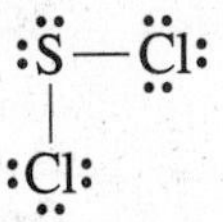 :S̈—C̈l: two single covalent bonds, three lone
 | pairs on each Cl, two lone pairs on S
:Cl:

c. 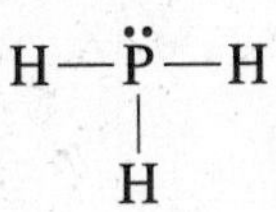 H—P̈—H   three single covalent bonds, one lone
 | pair on P, no lone pairs on H
 H

d. 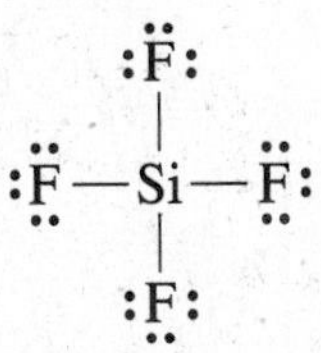   :F̈:   four single covalent bonds, three lone
 | pairs on each F, no lone pairs on Si
:F̈—Si—F̈:
 |
:F:

2. a. triple bond, one sigma and two pi
b. single bond, sigma
c. between H and C, single bond, sigma; between C and N,
triple bond, one sigma and two pi
d. between C and each O, double bond, one sigma and one pi
for each double bond

3. The bond dissociation energy and bond strength of the bond between the double-bonded N atoms would be expected to be greater than that between the single-bonded C atoms.

4. The bond length, strength, and dissociation energy of the triple-bonded C atoms would be expected to be greater than those of the double-bonded C atoms.

5. exothermic

6. **a.** sulfur dioxide
   b. tetraphosphorus decoxide
   c. dinitrogen trioxide
   d. silicon hexafluoride

7. **a.** carbonic acid
   b. hydriodic acid
   c. perchloric acid
   d. sulfurous acid

8. **a.** S_2Cl_2
   b. N_2O_4
   c. H_2S
   d. H_2SO_4

9. **a.**

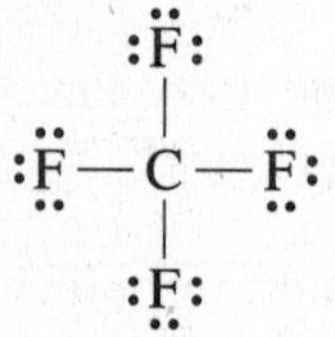

$$\ddot{\underset{\cdot\cdot}{F}} - \overset{\displaystyle :\ddot{F}:}{\underset{\displaystyle :\ddot{F}:}{C}} - \ddot{\underset{\cdot\cdot}{F}}:$$

   b. $:C\equiv O:$

   c. $:\!\overset{\cdot\cdot}{S}\!=\!Si\!=\!\overset{\cdot\cdot}{S}\!:$

   d.
$$\left[\; \overset{\displaystyle H}{\underset{\displaystyle H}{H-N-H}} \;\right]^{+}$$

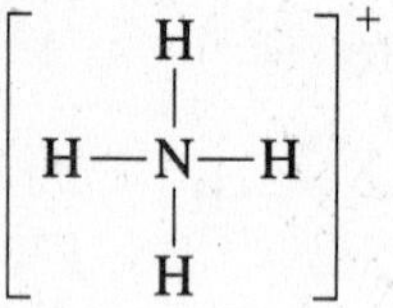

10.

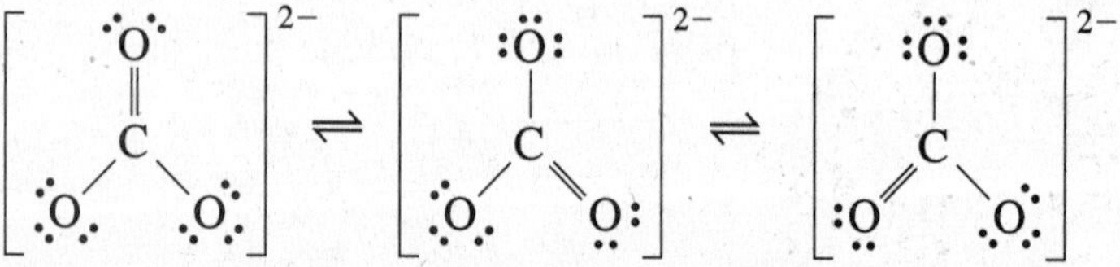

11. **a.** There is an odd number of valence electrons.
    b. Boron has fewer than an octet of electrons.
    c. Phosphorus has more than an octet of electrons.

12. **a.** four sp^3 hybrid orbitals, bent
    b. two sp hybrid orbitals, linear

   c. four sp^3 hybrid orbitals, tetrahedral

   d. four sp^3 hybrid orbitals, trigonal pyramidal

   e. three sp^2 hybrid orbitals, trigonal planar

13. a. polar bonds, tetrahedral, nonpolar molecule

   b. polar bonds, trigonal pyramidal, polar molecule

   c. polar bonds, bent, polar molecule

   d. polar bonds, tetrahedral, polar molecule

Chapter 9 Review

14. Atoms bond in order to achieve full outer energy levels, like those characteristic of the noble gases.

15. A covalent bond is one in which electrons are shared.

16. A sigma bond is one in which the electron pair is shared in an area centered between the two atoms. A pi bond is one in which parallel orbitals overlap, such that the shared electrons occupy space above and below a line between the two atoms. One sigma bond and two pi bonds are found in a triple bond.

17. Bond dissociation energy is the energy required to break a covalent bond.

18. endothermic reaction

19. Name the first element in the formula first, using its element name, unchanged. Name the second element, using the root of its name and adding the suffix *-ide*. Use prefixes to indicate the number of each type of atom present.

20. Predict the location of atoms as terminal or central. Find the total number of electrons available for bonding, subtracting or adding the charge for positive or negative ions, respectively. Divide by 2 to obtain the number of bonding pairs. Place one bonding pair between the central atom and each terminal atom. Subtract the number of bonding pairs used from the total number of bonding pairs obtained earlier. Place lone pairs around the terminal atoms to satisfy the octet rule. Assign remaining pairs to the central atom. If that atom is not surrounded by four electron pairs, convert one or two lone pairs on the terminal atoms to a double or triple bond to the central atom.

21. Resonance occurs when more than one valid Lewis structure can be written for a molecule or an ion.

22. The VSEPR model is used to determine molecular shape.

23. Hybridization is a process in which atomic orbitals are mixed to form new, identical hybrid orbitals.

24. A bond is polar when there is an electronegativity difference between the two bonded atoms. A molecule is polar when there are polar bonds and the molecular shape is not symmetric.

Chapter 10

Practice Problems

1. a. lithium(s) + chlorine(g) $\rightarrow$ lithium chloride(s)
$Li(s) + Cl_2(g) \rightarrow LiCl(s)$

b. nitrogen(g) + oxygen(g) $\rightarrow$ nitrogen dioxide(g)
$N_2(g) + O_2(g) \rightarrow NO_2(g)$

c. iron(s) + copper(II) nitrate(aq) $\rightarrow$ copper(s) + iron(II) nitrate(aq)
$Fe(s) + Cu(NO_3)_2(aq) \rightarrow Cu(s) + Fe(NO_3)_2(aq)$

2. a. $2K(s) + 2H_2O(l) \rightarrow H_2(g) + 2KOH(aq)$

b. $CaCl_2(aq) + Na_2CO_3(aq) \rightarrow CaCO_3(s) + 2NaCl(aq)$

c. $Br_2(l) + 2LiI(s) \rightarrow 2LiBr(s) + I_2(s)$

3. a. synthesis; $N_2(g) + 3H_2(g) \rightarrow 2NH_3(g)$

b. decomposition; already balanced

c. synthesis and combustion; $2Se(s) + 3O_2(g) \rightarrow 2SeO_3(g)$

d. combustion; $C_2H_4(g) + 3O_2(g) \rightarrow 2CO_2(g) + 2H_2O(g)$

4. a. yes; $Cl_2(g) + 2KI(aq) \rightarrow 2KCl(aq) + I_2(s)$

b. yes; $Mg(s) + CuSO_4(aq) \rightarrow MgSO_4(aq) + Cu(s)$

c. no

d. yes; $Pb(s) + 2AgNO_3(aq) \rightarrow Pb(NO_3)_2(aq) + 2Ag(s)$

5. a. $ZnBr_2(aq) + 2KOH(aq) \rightarrow 2KBr(aq) + Zn(OH)_2(s)$

b. $CuSO_4(aq) + BaCl_2(aq) \rightarrow CuCl_2(aq) + BaSO_4(s)$

c. $2Fe(NO_3)_3(aq) + 3Na_2CO_3(aq) \rightarrow$
$$Fe_2(CO_3)_3(s) + 6NaNO_3(aq)$$

6. a. chemical: $Pb(NO_3)_2(aq) + 2NH_4Cl(aq) \rightarrow$
$$2NH_4NO_3(aq) + PbCl_2(s)$$
complete ionic: $Pb^{2+}(aq) + 2NO_3^-(aq) + 2NH_4^+(aq) +$
$2Cl^-(aq) \rightarrow 2NH_4^+(aq) + 2NO_3^-(aq) + PbCl_2(s)$
net ionic: $Pb^{2+}(aq) + 2Cl^-(aq) \rightarrow PbCl_2(s)$

b. chemical: $2AlCl_3(aq) + 3Na_2CO_3(aq) \rightarrow$
$$6NaCl(aq) + Al_2(CO_3)_3(s)$$
complete ionic: $2Al^{3+}(aq) + 6Cl^-(aq) + 6Na^+(aq) +$
$3CO_3^{2-}(aq) \rightarrow 6Na^+(aq) + 6Cl^-(aq) + Al_2(CO_3)_3(s)$
net ionic: $2Al^{3+}(aq) + 3CO_3^{2-}(aq) \rightarrow Al_2(CO_3)_3(s)$

7. a. chemical: $2HNO_3(aq) + Ba(OH)_2(aq) \rightarrow$
$$Ba(NO_3)_2(aq) + 2H_2O(l)$$
complete ionic: $2H^+(aq) + 2NO_3^-(aq) + Ba^{2+}(aq) +$
$2OH^-(aq) \rightarrow Ba^{2+}(aq) + 2NO_3^-(aq) + 2H_2O(l)$
net ionic: $2H^+(aq) + 2OH^-(aq) \rightarrow 2H_2O(l)$,
or, with coefficients reduced to lowest terms,
$H^+(aq) + OH^-(aq) \rightarrow H_2O(l)$

b. chemical: $H_2SO_4(aq) + 2NaOH(aq)$
$$\rightarrow Na_2SO_4(aq) + 2H_2O(l)$$
complete ionic: $2H^+(aq) + SO_4^{2-}(aq) + 2Na^+(aq) +$
$2OH^-(aq) \rightarrow 2Na^+(aq) + SO_4^{2-}(aq) + 2H_2O(l)$
net ionic: $2H^+(aq) + 2OH^-(aq) \rightarrow 2H_2O(l)$,
or, with coefficients reduced to lowest terms,
$H^+(aq) + OH^-(aq) \rightarrow H_2O(l)$

c. chemical: $H_3PO_4(aq) + 3LiOH(aq)$
$$\rightarrow Li_3PO_4(aq) + 3H_2O(l)$$
complete ionic: $3H^+(aq) + PO_4^{3-}(aq) + 3Li^+(aq) +$
$3OH^-(aq) \rightarrow 3Li^+(aq) + PO_4^{3-}(aq) + 3H_2O(l)$
net ionic: $3H^+(aq) + 3OH^-(aq) \rightarrow 3H_2O(l)$,
or, with coefficients reduced to lowest terms,
$H^+(aq) + OH^-(aq) \rightarrow H_2O(l)$

8. a. chemical: $HCl(aq) + NaCN(aq) \rightarrow NaCl(aq) + HCN(g)$
complete ionic: $H^+(aq) + Cl^-(aq) + Na^+(aq) + CN^-(aq)$
$$\rightarrow Na^+(aq) + Cl^-(aq) + HCN(g)$$
net ionic: $H^+(aq) + CN^-(aq) \rightarrow HCN(g)$

b. chemical: $H_2SO_4(aq) + Rb_2S(aq) \rightarrow Rb_2SO_4(aq) + H_2S(g)$
complete ionic: $2H^+(aq) + SO_4{}^{2-}(aq) + 2Rb^+(aq) +$
$S^{2-}(aq) \rightarrow 2Rb^+(aq) + SO_4{}^{2-}(aq) + H_2S(g)$
net ionic: $2H^+(aq) + S^{2-}(aq) \rightarrow H_2S(g)$

Chapter 10 Review

9. A chemical reaction is a process in which the atoms of one or more substances are rearranged to form different substances. The reactants are shown to the left of the arrow, the products to the right. Physical states are shown by means of symbols in parentheses.

10. All of the equations identify the reactants and products, use plus signs to separate the reactants and products, use an arrow to indicate "yield," and use symbols to indicate the physical states of the substances. Word equations show the names of the substances. Skeleton equations show the chemical formulas of the substances but are not balanced in terms of numbers of atoms. Balanced chemical equations also show chemical formulas but use coefficients to balance the number of atoms of each element on both sides of the equation.

11. Write the skeleton equation. Count the atoms of the elements in the reactants and in the products. Change coefficients to make the number of atoms of each element equal on both sides of the equation. Write the coefficients in their lowest possible ratio. Check your work.

12. In synthesis reactions, two or more substances react to produce a single product. In combustion reactions, oxygen combines with a substance and energy is released as light and heat. In decomposition reactions, a single compound breaks down into two or more elements or new compounds.

13. In single-replacement reactions, atoms of one element replace the atoms of another element in a compound. In double-replacement reactions, there is an exchange of positive ions between two compounds.

14. An aqueous solution is a solution in which the solvent is water. A solid that is formed when two aqueous solutions are mixed is called a precipitate.

15. A complete ionic equation includes all of the particles in a solution, including spectator ions. A net ionic equation does not include the spectator ions.

16. Double-replacement reactions may form a precipitate, water, or a gas.

Chapter 11

Practice Problems

1. 9.45×10^{24} molecules CO_2

2. 3.27×10^{22} molecules H_2O

3. 1.53 mol Fe

4. 0.0341 mol sucrose

5. 1.08×10^{21} formula units $CuSO_4$

6. 839 g Sb

7. 5.53 g Se

8. 6.95 mol S

9. 12.5 mol He

10. 0.0213 mol Ni

11. 6.17×10^{21} atoms Pt

12. 1.88×10^{28} atoms S

13. 39.7 g Hg

14. 6.35×10^{-3} g

15. 4.36×10^{16} Pb ions

16. 0.220 mol C_6H_6

17. 2.856 mol $KClO_3$

18. 40.2 g $SnSO_4$

19. 3.40×10^2 g (340 g expressed to 3 significant figures)

20. 13.1 g Zn and 37.5 g $Cu(NO_3)_2$

21. 52.93% Al, 47.07% O

22. 16.39% Mg, 18.89% N, 64.72% O

23. 39.17% O

24. 49.30% N

25. The percent compositions will be the same because the ratio of carbon atoms to hydrogen atoms is the same, 1:1.

26. CH_2O

27. ZnN_2O_6

28. Al_4C_3

29. $C_3H_4O_3$

30. $C_8H_8N_2O_2$

31. $B_3H_6N_3$

32. $Ag_2C_2O_4$

33. P_2S_5; same as empirical formula

34. $C_9H_{12}N_6$

35. $CeI_3 \cdot 9H_2O$

36. $Co(NO_3)_2 \cdot 6H_2O$

38. $ZnSO_4 \cdot 7H_2O$

37. $Pb(C_2H_3O_2)_2 \cdot 3H_2O$

Chapter 11 Review

39. One mole of a substance consists of 6.02×10^{23} particles of that substance. The mass in grams of a mole of a substance is numerically equal to the sum of the atomic masses of the atoms that make up the particles of that substance.

40. The quantities of reactants and products in a chemical equation are given as numbers of particles or moles of particles, not as masses.

41. The molar mass of a compound in grams is numerically equal to the sum of the atomic masses of the atoms in a representative particle of the compound.

42. 28.014 g, which is twice the atomic mass expressed in grams

43. Yes; you would expect the compositions to be the same. A compound is a substance, and the law of definite proportions states that the composition of a substance is the same wherever the substance is found.

44. The molecular formula will be a whole-number multiple of the empirical formula but could also be the same as the empirical formula.

45. The six water molecules are incorporated in the crystal structure but are separate from the compound. The formula with water molecules shown separately better represents the composition of the compound.

Chapter 12

Practice Problems

1. a. $2H_2O_2(l) \rightarrow O_2(g) + 2H_2O(l)$

2 molecules $H_2O_2 \rightarrow$ 1 molecule O_2 + 2 molecules H_2O

2 moles $H_2O_2 \rightarrow$ 1 mole O_2 + 2 moles H_2O

$68.04 \text{ g } H_2O_2 = 32.00 \text{ g } O_2 + 36.04 \text{ g } H_2O$, as shown below.

$68.04 \text{ g reactant} = 68.04 \text{ g products}$

$$2 \text{ mol } H_2O_2 \times \frac{34.02 \text{ g } H_2O_2}{1 \text{ mol } H_2O_2} = 68.04 \text{ g } H_2O_2$$

$$1 \text{ mol } O_2 \times \frac{32.00 \text{ g } O_2}{1 \text{ mol } O_2} = 32.00 \text{ g } O_2$$

$$2 \text{ mol } H_2O \times \frac{18.02 \text{ g } H_2O}{1 \text{ mol } H_2O} = 36.04 \text{ g } H_2O$$

b. $H_2CO_3(aq) \rightarrow H_2O(l) + CO_2(g)$

1 formula unit $H_2CO_3 \rightarrow$ 1 molecule H_2O + 1 molecule CO_2

1 mole $H_2CO_3 \rightarrow$ 1 mole H_2O + 1 mole CO_2

$62.03 \text{ g } H_2CO_3 = 18.02 \text{ g } H_2O + 44.01 \text{ g } CO_2$, as shown below.

$62.03 \text{ g reactant} = 62.03 \text{ g products}$

$$1 \text{ mol } H_2CO_3 \times \frac{62.03 \text{ g } H_2CO_3}{1 \text{ mol } H_2CO_3} = 62.03 \text{ g } H_2CO_3$$

$$1 \text{ mol } H_2O \times \frac{18.02 \text{ g } H_2O}{1 \text{ mol } H_2O} = 18.02 \text{ g } H_2O$$

$$1 \text{ mol } CO_2 \times \frac{44.01 \text{ g } CO_2}{1 \text{ mol } CO_2} = 44.01 \text{ g } CO_2$$

c. $4HCl(aq) + O_2(g) \rightarrow 2H_2O(l) + 2Cl_2(g)$

4 molecules HCl + 1 molecule $O_2 \rightarrow$
2 molecules H_2O + 2 molecules Cl_2

4 mole HCl + 1 mol $O_2 \rightarrow$ 2 mole H_2O + 2 mole Cl_2

$146 \text{ g HCl} + 32.00 \text{ g } O_2 = 36.04 \text{ g } H_2O + 142 \text{ g } Cl_2$, as shown below.

$177.84 \text{ g reactants} = 177.84 \text{ g products}$

$$4 \text{ mol } HCl \times \frac{36.46 \text{ g } HCl}{1 \text{ mol } HCl} = 145.84 \text{ g } HCl$$

$$1 \text{ mol } O_2 \times \frac{32.00 \text{ g } O_2}{1 \text{ mol } O_2} = 32.00 \text{ g } O_2$$

$$2 \text{ mol } H_2O \times \frac{18.02 \text{ g } H_2O}{1 \text{ mol } H_2O} = 36.04 \text{ g } H_2O$$

$$2 \text{ mol } Cl_2 \times \frac{70.90 \text{ g } Cl_2}{1 \text{ mol } Cl_2} = 141.80 \text{ g } Cl_2$$

2. a. $N_2(g) + O_2(g) \rightarrow 2NO(g)$

$$\frac{1 \text{ mol } N_2}{1 \text{ mol } O_2}; \frac{1 \text{ mol } N_2}{2 \text{ mol } NO}; \frac{1 \text{ mol } O_2}{1 \text{ mol } N_2}; \frac{1 \text{ mol } O_2}{2 \text{ mol } NO};$$

$$\frac{2 \text{ mol } NO}{1 \text{ mol } N_2}; \frac{2 \text{ mol } NO}{1 \text{ mol } O_2}$$

b. $4NH_3(aq) + 5O_2(g) \rightarrow 4NO(g) + 6H_2O(l)$

$$\frac{4 \text{ mol } NH_3}{5 \text{ mol } O_2}, \frac{4 \text{ mol } NH_3}{4 \text{ mol } NO}, \frac{4 \text{ mol } NH_3}{6 \text{ mol } H_2O}, \frac{5 \text{ mol } O_2}{4 \text{ mol } NH_3},$$

$$\frac{5 \text{ mol } O_2}{4 \text{ mol } NO}, \frac{5 \text{ mol } O_2}{6 \text{ mol } H_2O}, \frac{4 \text{ mol } NO}{4 \text{ mol } NH_3}, \frac{4 \text{ mol } NO}{5 \text{ mol } O_2},$$

$$\frac{4 \text{ mol } NO}{6 \text{ mol } H_2O}, \frac{6 \text{ mol } H_2O}{4 \text{ mol } NH_3}, \frac{6 \text{ mol } H_2O}{5 \text{ mol } O_2}, \frac{6 \text{ mol } H_2O}{4 \text{ mol } NO}$$

c. $4HCl(aq) + O_2(g) \rightarrow 2H_2O(l) + 2Cl_2$

$$\frac{4 \text{ mol } HCl}{1 \text{ mol } O_2}, \frac{4 \text{ mol } HCl}{2 \text{ mol } H_2O}, \frac{4 \text{ mol } HCl}{2 \text{ mol } Cl_2}, \frac{1 \text{ mol } O_2}{4 \text{ mol } HCl},$$

$$\frac{1 \text{ mol } O_2}{2 \text{ mol } Cl_2}, \frac{1 \text{ mol } O_2}{2 \text{ mol } H_2O}, \frac{2 \text{ mol } H_2O}{4 \text{ mol } HCl}, \frac{2 \text{ mol } H_2O}{1 \text{ mol } O_2},$$

$$\frac{2 \text{ mol } H_2O}{2 \text{ mol } Cl_2}, \frac{2 \text{ mol } Cl_2}{4 \text{ mol } HCl}, \frac{2 \text{ mol } Cl_2}{1 \text{ mol } O_2}, \frac{2 \text{ mol } Cl_2}{2 \text{ mol } H_2O}$$

3. 240 mol LiOH is needed.

4. $2KClO_3(s) \rightarrow 2KCl(s) + 3O_2(g)$

a. 15 mol O_2

b. 3 mol KCl

c. 33 mol $KClO_3$, or 30 mol $KClO_3$ using significant figures

5. From the reaction $2Na(s) + Cl_2(g) \rightarrow 2NaCl(s)$, 321 g NaCl is produced.

6. a. $2NaI(aq) + Cl_2(g) \rightarrow 2NaCl(aq) + I_2(s)$
  b. 213 g Cl_2

7. a. $Zn(s) + 2HCl(aq) \rightarrow ZnCl_2(aq) + H_2(g)$
  b. 365 g HCl

8. a. $Mg(s) + 2HCl(aq) \rightarrow MgCl_2(aq) + H_2(g)$; 157 g $MgCl_2$
  b. $Cu(s) + 2AgNO_3(aq) \rightarrow Cu(NO_3)_2(aq) + 2Ag(s)$;
    2.24 g Cu
  c. $2NaCl(aq) + H_2SO_4(aq) \rightarrow Na_2SO_4 + 2HCl(g)$; 9.36 g HCl
  d. $3AgCH_3COO(aq) + Na_3PO_4(aq) \rightarrow Ag_3PO_4(s) +$
    $3NaCH_3COO(aq)$; 25.1 g Ag_3PO_4

9. a. N_2
  b. 1.22×10^3 g NH_3
  c. 34 g H_2

10. a. $2Al(s) + 3Cl_2(g) \rightarrow 2AlCl_3(s)$
  b. Cl_2
  c. 6.8 g $AlCl_3$
  d. 1.8 g Al

11. a. 80%
  b. 76%
  c. 94%

12. a. 19 g
  b. 66 g
  c. 5.7 g

13. theoretical yield: 39.2 g; percent yield: 75.3%

Chapter 12 Review

14. Stoichiometry is the study of quantitative relationships between amounts of reactants used and products formed by a chemical reaction.

15. Sample questions: What mass of HCl is needed to react completely with a known mass of O_2? How much water will be produced if a given mass of HCl is used in the reaction?

16. The law of conservation of mass states that matter is neither created nor destroyed; thus, in a chemical reaction, the mass of the reactants must equal the mass of the products.

17. The limiting reactant limits the amount of product that can form from the reaction. An excess reactant is left over after the reaction is complete and all of the limiting reactant has been used up.

Chapter 13

Practice Problems

1. He diffuses about 2.25 times faster than Ne does.

2. 1.61; ammonia diffuses more rapidly than carbon dioxide does.

3. 2.36; Ar diffuses more rapidly than Rn does.

4. 50 g/mol

5. 0.24 atm

6. 11.0 kPa

7. 38.4 kPa

8. 40.7 kPa

9. a. ionic
   b. covalent network
   c. metallic
   d. molecular

10. a. sublimation
   b. melting
   c. vaporization
   d. sublimation

11. a. freezing
   b. condensation
   c. condensation
   d. deposition
   e. condensation

12. a. The ice would melt (become liquid water) and vaporize (become water vapor).
   b. 1.00–217.75 atm
   c. Ice, liquid water, and water vapor coexist at the triple point.

13. a. The dry ice sublimes to carbon dioxide gas.

    b. The dry ice would melt (become a liquid) and vaporize (become a gas.)

Chapter 13 Review

14. The volume of gas particles is small compared with the volume of empty space between the particles. Gas particles are in constant, random motion, and collisions between them are elastic. At a given temperature, all gases have the same average kinetic energy.

15. Oxygen gas has a lower density than silver because the oxygen particles are spaced farther apart than the silver particles.

16. Diffusion is the movement of one material through another. Effusion is the escape of a gas through a small opening in its container.

17. Fastest: He; slowest: Xe; rate of diffusion is inversely proportional to molar mass.

18. No; the particles will have different kinetic energies because their velocities differ.

19. The container of refrigerated syrup will take the longest to pour because viscosity increases as temperature decreases.

20. Properties of liquids include the following: have a definite volume, conform to the shape of their container, are denser than most gases, have fluidity, are relatively incompressible, diffuse, have surface tension, and show capillary action.

21. A concave surface forms when the attraction between the liquid particles and glass particles (adhesion) is greater than the attraction between the liquid particles (cohesion). A convex surface forms when the attraction between the liquid particles is greater than the attraction between the liquid particles and the glass particles.

22. There are no significant attractive forces between gas particles, and most gas particles have a smaller mass than most liquid particles.

23. The strong attractive forces between the solid particles keeps them closely bound in fixed positions.

24. The particles of an amorphous solid are arranged randomly; the particles of a crystalline solid are arranged in an orderly, geometric, three-dimensional lattice.

25. a melting

    b. sublimation

    c. freezing

    d. vaporization

    e. condensation

    f. deposition

Chapter 14

Practice Problems

1. 126 kPa

2. 262.2 mL

3. 97.7 L

4. 10 500 L

5. 46°C (319 K)

6. 5.94 L

7. 18°C (291 K)

8. 8680 L

9. 1.44 L

10. 95.6 kPa

11. 1.00×10^2 kPa

12. 0.145 mol

13. 9.18 L

14. 6420 L

15. 367 g

16. 5.45×10^{-4} mol

17. 4.86 atm

18. 1.2 L

19. 0.0102 mol or 1.02×10^{-2} mol

20. Molar mass = 40.0 g/mol. The gas is likely to be argon.

21. Molar mass = 64.1 g/mol. The gas must be SO_2, which has a molar mass of 64.06 g/mol.

22. 33.6 L

23. 282 K or 9°C

24. 5.99 L

25. 3.69 L

26. 24.0 g

27. $2KClO_3(s) \rightarrow 2KCl_2(s) + 3O_2(g)$; 1.77 g

Chapter 14 Review

28. The spacing is much greater than the size of the particles. No kinetic energy is lost in the collisions.

29. Squeezing concentrates the gas particles in a smaller space. As a result, they collide with each other and the walls of their container more often.

30. The volume is reduced to one-third the original volume.

31. The volume doubles.

32. 44.6 mol

33. 240 mL

Chapter 15

Practice Problems

1. a. 61 g $CaCl_2$/100 g H_2O
   b. 90 g $CaCl_2$/100 g H_2O
   c. 5 g $Ce_2(SO_4)_3$
   d. 1 g $Ce_2(SO_4)_3$
   e. 60 g $KClO_3$

2. 94 atm

3. 2.1 atm

4. 0.852%

5. 7.35 g

6. 492.65 g

7. 26%

8. 0.868 L

9. 0.328M

10. 1.09M

11. 326 mL

12. 0.802M

13. 0.350m

14. 0.0300m

15. a. water, 0.830; ethanol, 0.170
   b. water, 0.9853; calcium chloride, 0.0147
   c. carbon tetrachloride, 0.644; benzene, 0.356

16. BP = 102.06°C; FP = −7.48 °C

17. 2.33m

18. a. suspension
   b. solution or colloid
   c. colloid
   d. solution

Chapter 15 Review

19. Solvation is a process in which solvent particles surround solute particles to form a solution. A substance dissolves when attractive forces between solvent and solute particles are strong enough to overcome the attractive forces holding the solute particles together.

20. A saturated solution contains the maximum amount of dissolved solute for a given amount of solvent at a specific temperature and pressure. An unsaturated solution contains less dissolved solute for a given temperature and pressure than does a saturated solution. A supersaturated solution contains more dissolved solute than does a saturated solution at the same temperature.

21. Henry's law states that at a given temperature, the solubility (S) of a gas in a liquid is directly proportional to the pressure (P) of the gas above the liquid; $S_1/P_1 = S_2/P_2$

22. Percent by mass is calculated by dividing the mass of solute by the mass of solution, then multiplying the value by 100.

23. A $2M$ solution is 2 molar and contains 2 moles of solute per *liter* of solution. A $2m$ solution is 2 molal and contains 2 moles of solute per *kilogram* of solvent.

24. A colligative property is a physical property of a solution that is affected by the number of solute particles but not by the identity of those particles. Colligative properties include vapor pressure lowering, osmotic pressure, boiling point elevation, and freezing point depression.

25. No; the freezing point depression of the sodium chloride solution would be about twice as great as that of the sucrose solution because sodium chloride is ionic and dissociates to produce two particles per formula unit, unlike the nonelectrolyte sucrose.

26. A solution is a homogeneous mixture whose solute particles are less than 1 nm in diameter and remain distributed throughout in a single phase. Solutions do not show the Tyndall effect. A suspension is a heterogeneous mixture whose suspended particles are greater than 1000 nm in diameter and eventually settle out.

A colloid is a heterogeneous mixture whose particles are between 1 nm and 1000 nm in diameter and normally do not settle out. Colloids show the Tyndall effect.

Chapter 16

Practice Problems

1. 1674 J or 400 cal

2. Specific heat = 0.898 J/(g·°C). The specific heat is very close to the value for aluminum.

3. 44.3°C

4. 112°C

5. 0.40 J/(g·°C)

6. 47.9°C

7. **a.** 10.7 kJ
   b. 83.8 kJ

   The heat needed for boiling is much larger than the heat needed for melting.

8. 833 kJ is evolved.

9. 239 g

10. $\Delta H_{rxn} = -112$ kJ

11. $\Delta H_{rxn} = -129$ kJ

12. **a.** +40.9 kJ
    b. −153.87 kJ
    c. −1322.91 kJ
    d. +3.0 kJ

13. **a.** negative
    b. positive
    c. positive
    d. positive
    e. negative

14. **a.** 175 kJ, nonspontaneous
    b. −8 kJ, spontaneous
    c. 27 kJ, nonspontaneous

15. 391 K, or 118°C

Chapter 16 Review

16. Use the equation for the amount of heat absorbed.

$$q = c \times m \times \Delta T$$

Both samples have the same mass and the same temperature change, so the sample with the higher specific heat absorbs more energy as it is heated. From Table 16-2 in your textbook, the specific heat of ethanol is 2.44 J/(g·°C), whereas that of aluminum is 0.897 J/(g·°C), so the ethanol absorbs more energy.

17. The universe is defined as the system plus the surroundings. The system is the reaction or process being studied. The surroundings include everything in the universe other than the system.

18. Melting of a solid and boiling of a liquid are endothermic processes, so ΔH is positive. Condensation of a vapor and solidification of a liquid are exothermic, so ΔH is negative.

19. The standard heat of formation (ΔH_f°) of ethane gas is the enthalpy change that accompanies the formation of one mole of ethane in its standard state from its constituent elements, carbon and hydrogen, in their standard states. The standard state of a substance is its normal state at one atmosphere pressure and 298 K (25°C). The equation for the formation of ethane is as follows.

$$2C(s) + 3H_2(g) \rightarrow C_2H_6(g)$$

20. $\Delta G_{system} = \Delta H_{system} - T\Delta S_{system}$

a. $\Delta G_{system} = (\text{positive term}) - T(\text{negative term})$

$\Delta G_{system} = (\text{positive term}) + (\text{positive term})$

$\Delta G_{system} > 0$ not spontaneous

b. $\Delta G_{system} = (\text{negative term}) - T(\text{negative term})$

$\Delta G_{system} = (\text{negative term}) + (\text{positive term})$

The sign of ΔG is uncertain, so spontaneity is uncertain.

c. $\Delta G_{system} = (\text{negative term}) - T(\text{positive term})$

$\Delta G_{system} = (\text{negative term}) + (\text{negative term})$

$\Delta G_{system} < 0$ spontaneous

Chapter 17

Practice Problems

1. $0.000\ 032\ 7\ \dfrac{\text{mol}}{\text{L·s}}$ or $3.27 \times 10^{-5}\ \dfrac{\text{mol}}{\text{L·s}}$

2. $0.000\ 065\ 3\ \dfrac{\text{mol}}{\text{L·s}}$, or $6.53 \times 10^{-5}\ \dfrac{\text{mol}}{\text{L·s}}$

3. The rate in problem 2 is twice the rate in problem 1 because two moles of HBr are produced for every one mole of Br_2 or HCOOH that reacts.

4. Increasing the pressure increases the concentration by increasing the number of particles in a given space.

5. Adding more gas also increases the concentration by adding particles to a given space.

6. The small scraps of kindling provide more surface area for combustion. Therefore, they burn faster than the logs.

7. Cooling the flask reduces the average kinetic energy of the reacting particles, reducing the frequency of collisions and slowing the reaction.

8. The powdered palladium acts as a catalyst, providing a surface on which the molecules can combine with oxygen at a lower activation energy (a lower temperature).

9. The rate triples. The rate is halved.

10. The rate doubles. The rate increases to nine times the original rate because the rate depends on $[Y]^2$.

11. The reaction is second order with respect to each reactant. Therefore, the rate equation is rate $= k[D]^2[E]^2$.

Chapter 17 Review

12. There are three possibilities. Measure the reduction in concentration of H_2, the reduction in concentration of I_2, or the increase in concentration of HI. In all three cases, time is the second quantity that must be measured.

13. $1.22 \times 10^{-4} \dfrac{mol}{L \cdot s}$

14. The particles must collide in the correct orientation and with sufficient energy to form an activated complex.

15. Increasing reactant concentration increases the frequency of collisions, thereby increasing the rate of formation of activated complexes.

16. The reaction rate is quadrupled because the rate depends on the square of the concentration of Q.

17. The rate-determining step is the slowest step in the sequence. Therefore, the reaction can proceed no faster than that step.

Chapter 18

Practice Problems

1. a. $K_{eq} = \dfrac{[CH_4][CO]}{[C_2H_4O]}$ **c.** $K_{eq} = \dfrac{[NO]^4}{[N_2O]^2[O_2]}$

b. $K_{eq} = \dfrac{[O_3]^2}{[O_2]^3}$ **d.** $K_{eq} = \dfrac{[N_2]^2[H_2O]^6}{[NH_3]^4[O_2]^3}$

2. a. $K_{eq} = [C_4H_{10}(g)]$
b. $K_{eq} = [NH_3][H_2S]$

c. $K_{eq} = \dfrac{[CO_2]}{[CO]}$

d. $K_{eq} = [NH_3]^2[CO_2][H_2O]$

3. 4.21

4. 0.257

5. Substitute the concentrations into the equilibrium constant expression, $K_{eq} = \dfrac{[C]^2}{[A][B]^3}$. The result is 28.0 for the first reaction mixture and 9.36 for the second, so the first mixture is not at equilibrium.

6. a. shift to the left   **b.** shift to the right   **c.** shift to the left

7. a. shift to the right   **c.** no shift
b. shift to the left   **d.** shift to the left

8. a. increase the temperature   **c.** decrease the temperature
b. increase the temperature   **d.** increase the temperature

9. Decreasing the temperature causes a shift to the left, while decreasing the pressure causes a shift to the right, so the overall effect is uncertain.

10. **a.** 0.189 mol/L   **b.** 0.486 mol/L   **c.** 0.156 mol/L

11. $[NO_2] = 0.042$ mol/L; $[N_2O_4] = 0.063$ mol/L

12. **a.** 2.6×10^{-3} mol/L   **c.** 1.4×10^{-2} mol/L
    b. 9.9×10^{-11} mol/L   **d.** 3.6×10^{-9} mol/L

13. **a.** 1.1×10^{-5} mol/L   **c.** 3.6×10^{-2} mol/L
    b. 1.2×10^{-5} mol/L   **d.** 1.9×10^{-5} mol/L

14. **a.** For $BaSO_4$: Q_{sp} $(1.2 \times 10^{-4}) > K_{sp}$ (1.1×10^{-10})
    A precipitate forms.

    b. For $PbBr_2$: Q_{sp} $(7.5 \times 10^{-7}) < K_{sp}$ (6.6×10^{-6})
    No precipitate forms.

    c. For Ag_2CrO_4: Q_{sp} $(4.7 \times 10^{-9}) > K_{sp}$ (1.1×10^{-12})
    A precipitate forms.

15. For $CaSO_4$: Q_{sp} $(3.3 \times 10^{-5}) < K_{sp}$ (4.9×10^{-5})
    No precipitate forms.

Chapter 18 Review

16. Products are favored in this equilibrium, so it is likely that K_{eq} is greater than 1.

17. $K_{eq} = \dfrac{[C]^2}{[A]^2[B]} = \dfrac{(3.0 \times [A])^2}{[A]^2[0.14]} = \dfrac{9.0 \times [A]^2}{[A]^2[0.14]} = \dfrac{9.0}{0.14} = 64$

18. Yes; when a stress is placed on an equilibrium system, the system shifts to relieve the stress. Examples of stresses include addition or removal of reactants or products, or a change in the volume of the reaction vessel. The equilibrium constant varies only with temperature, so if a stress is applied while the temperature remains constant, the equilibrium will shift, but K_{eq} will not change.

19. The equilibrium may be shifted toward the products by adding a reactant to the system, removing a product from the system, decreasing the volume of the reaction vessel, or increasing the pressure on the system.

20. The K_{sp} expressions are as follows.

For AgI: $K_{sp} = [Ag^+][I^-]$

For Fe(OH)$_2$: $K_{sp} = [Fe^{2+}][OH^-]^2$

For these sparingly soluble compounds, the ion concentrations are much less than $1M$. Note that K_{sp} for Fe(OH)$_2$ is the product of three of these very small ion concentration terms, whereas K_{sp} for AgI is the product of only two ion concentrations. Therefore, K_{sp} is smaller for Fe(OH)$_2$ than for AgI, even though Fe(OH)$_2$ has a higher molar solubility.

21. According to the common ion effect, the solubility of Ag_2CO_3 would be lower in a solution of Na_2CO_3 than in pure water because of the presence of the common ion, $CO_3{}^{2-}$, in the Na_2CO_3 solution.

Chapter 19

Practice Problems

1.

	Acid	Conjugate Base	Base	Conjugate Acid
a.	H_2SO_3	$HSO_3{}^-$	H_2O	H_3O^+
b.	H_2O	OH^-	$HPO_4{}^{2-}$	$H_2PO_4{}^-$
c.	$HSeO_3{}^-$	$SeO_3{}^{2-}$	H_2O	H_3O^+

2. a. $H_2CO_3(aq) + H_2O(l) \rightleftharpoons H_3O^+(aq) + HCO_3{}^-(aq)$

$HCO_3{}^-(aq) + H_2O(l) \rightleftharpoons H_3O^+(aq) + CO_3{}^{2-}(aq)$

b. $H_2CrO_4(aq) + H_2O(l) \rightleftharpoons H_3O^+(aq) + HCrO_4{}^-(aq)$

$HCrO_4{}^-(aq) + H_2O(l) \rightleftharpoons H_3O^+(aq) + CrO_4{}^{2-}(aq)$

3. a. $HF(aq) + H_2O(l) \rightleftharpoons H_3O^+(aq) + F^-(aq)$

$$K_a = \frac{[H_3O^+][F^-]}{[HF]}$$

b. $HBrO(aq) + H_2O(l) \rightleftharpoons H_3O^+(aq) + BrO^-(aq)$

$$K_a = \frac{[H_3O^+][BrO^-]}{[HBrO]}$$

4. $HSO_3^-(aq) + H_2O(l) \rightleftharpoons H_3O^+(aq) + SO_3^{2-}(aq)$

$$K_a = \frac{[H_3O^+][SO_3^{2-}]}{[HSO_3^-]}$$

5. a. $C_4H_9NH_2(aq) + H_2O(l) \rightleftharpoons C_4H_9NH_3^+(aq) + OH^-(aq)$

$$K_b = \frac{[C_4H_9NH_3^+][OH^-]}{[C_4H_9NH_2]}$$

b. $PO_4^{3-}(aq) + H_2O(l) \rightleftharpoons HPO_4^{2-}(aq) + OH^-(aq)$

$$K_b = \frac{[HPO_4^{2+}][OH^-]}{[PO_4^{3-}]}$$

c. $HCO_3^-(aq) + H_2O(l) \rightleftharpoons H_2CO_3(aq) + OH^-(aq)$

$$K_b = \frac{[H_2CO_3][OH^-]}{[HCO_3^-]}$$

6. a. $[H^+] = 1.0 \times 10^{-8}M$, basic
b. $[OH^-] = 1.0 \times 10^{-7}M$, neutral
c. $[OH^-] = 1.2 \times 10^{-12}M$, acidic

7. a. pH = 14.00; pOH = 0.00
b. pH = 6.75; pOH = 7.25
c. pH = 2.57; pOH = 11.43
d. pH = 12.79; pOH = 1.21

8. a. $[H^+] = 1.6 \times 10^{-3}M$; $[OH^-] = 6.3 \times 10^{-12}M$
b. $[H^+] = 6.5 \times 10^{-14}M$; $[OH^-] = 0.15M$
c. $[H^+] = 5.8 \times 10^{-6}M$; $[OH^-] = 1.7 \times 10^{-9}M$

9. a. 1.82   **c.** 3.60
b. 12.13   **d.** 11.90

10. a. 1.6×10^{-2} **b.** 6.4×10^{-5}

11. $0.531M$

12. $0.1234M$

13. $0.183M$

Chapter 19 Review

14. An Arrhenius base is a substance that contains a hydroxide (OH) group and dissociates in aqueous solution to produce hydroxide ions. A Brønsted-Lowry base is a hydrogen-ion acceptor that may or may not contain a hydroxide group.

15. Acid X is a weak acid that ionizes only partially in water. Acid Y has an even smaller K_a, so it is weaker than X.

16. A decrease in the hydroxide ion concentration is accompanied by an increase in the hydrogen ion concentration. The solution becomes more acidic, and the pH decreases.

17. The pH of solution A is 9.0, and the pH of solution B is $14.00 - pOH = 14.00 - 3.0 = 11.0$. Thus, both solutions are basic because their pH values are above 7. Solution A has a lower pH, so it has a higher concentration of hydrogen ions.

18. Find the hydrogen ion concentration of the HCl solution.
$[H^+] = $ antilog $(-pH) = $ antilog $(-2.00) = 1.0 \times 10^{-2}M = 0.010M$
HCl is a strong acid, so it ionizes completely and its molarity equals $[H^+]$. Thus, the solution is $0.010M$ HCl.

19. A large pH change occurs at the equivalence point of an acid-base titration. This pH change can be detected with a pH meter or an acid-base indicator.

20. A buffer is a mixture of a weak acid and its conjugate base or a weak base and its conjugate acid. Sodium formate dissociates in water to produce sodium ions and formate ions.
$$NaHCOO(s) \rightarrow Na^+(aq) + HCOO^-(aq)$$
The $HCOO^-$ ion is the conjugate base of formic acid, HCOOH, a weak acid. It is likely that the buffer solution consists of an aqueous solution of sodium formate and formic acid.

Chapter 20

Practice Problems

1. a. Zn is oxidized; Ni^{2+} is reduced.
　　oxidizing agent: Ni^{2+}
　　reducing agent: Zn

b. I^- is oxidized; Br is reduced.
oxidizing agent: Br_2
reducing agent: I^-

c. O is oxidized; N is reduced.
oxidizing agent: N
reducing agent: O

d. H is oxidized; S is reduced.
oxidizing agent: S_2
reducing agent: H_2

2. a. $+4$ **c.** -1 **e.** $+4$
    b. -4 **d.** $+6$ **f.** $+5$

3. a. $+5$ **c.** $+6$ **e.** $+5$
    b. $+1$ **d.** $+4$ **f.** $+4$

4. a. $2Cu_2O + 2NO \rightarrow 4CuO + N_2$
    b. $Al_2O_3 + 3C + N_2 \rightarrow 2AlN + 3CO$
    c. $3Ag + 4HNO_3 \rightarrow 3AgNO_3 + NO + 2H_2O$
    d. $I_2 + 5HClO + H_2O \rightarrow 2HIO_3 + 5HCl$

5. a. $2Al(s) + 3Ni^{2+}(aq) \rightarrow 2Al^{3+}(aq) + 3Ni(s)$
    b. $3HS^-(aq) + IO_3^-(aq) + 3H^+(aq) \rightarrow$
$$I^-(aq) + 3S(s) + 3H_2O(l)$$

c. $I_2(s) + 5HClO(aq) + H_2O(l) \rightarrow$
$$2IO_3^-(aq) + 5Cl^-(aq) + 7H^+(aq)$$

d. $3MnO_4^{2-}(aq) + 4H^+(aq) \rightarrow$
$$2MnO_4^-(aq) + MnO_2(s) + 2H_2O(l)$$

6. a. $I_2(s) + H_2SO_3(aq) + H_2O(l) \rightarrow$
$$2I^-(aq) + HSO_4^-(aq) + 3H^+(aq)$$

b. $5Fe^{2+}(aq) + MnO_4^-(aq) + 8H^+(aq) \rightarrow$
$$5Fe^{3+}(aq) + Mn^{2+}(aq) + 4H_2O(l)$$

c. $3Zn(s) + Cr_2O_7^{2-}(aq) + 14H^+(aq) \rightarrow$
$$3Zn^{2+}(aq) + 2Cr^{3+}(aq) + 7H_2O(l)$$

d. $IO_3^-(aq) + 5I^-(aq) + 6H^+(aq) \rightarrow 3I_2(s) + 3H_2O(l)$

Chapter 20 Review

7. Reactions b and c are double-replacement reactions that involve only the exchange of ions. No electrons are transferred between atoms, so the reactions are not redox reactions.

8. a. The oxidation number of bromine is -1. Bromine gains an electron when it is bonded to lithium because lithium has a low electronegativity.

b. The oxidation number of bromine is $+3$. Bromine loses electrons when it is bonded to fluorine because fluorine is highly electronegative.

c. The oxidation number of bromine is $+5$. Bromine loses electrons when it is bonded to oxygen because oxygen is highly electronegative.

9. Chlorine is a stronger oxidizing agent than iodine is because chlorine is more electronegative than iodine is and therefore gains electrons from other atoms more easily than iodine does.

10. The oxidation-number method or the half-reaction method can be used to obtain the balanced equation.

$$Fe_2O_3(s) + 3CO(g) \rightarrow 2Fe(l) + 3CO_2(g)$$

11. Chlorine is both oxidized and reduced in the reaction. Its oxidation number increases from $+5$ to $+7$ in oxidation and decreases from $+5$ to $+4$ in reduction. The half-reactions are as follows.

$$ClO_3^- \rightarrow ClO_4^- + 2e^- \text{ (oxidation)}$$
$$ClO_3^- + e^- \rightarrow ClO_2 \text{ (reduction)}$$

Balance the half-reactions. The reaction occurs in acid solution.

$$ClO_3^- + H_2O \rightarrow ClO_4^- + 2e^- + 2H^+ \text{ (oxidation)}$$
$$ClO_3^- + e^- + 2H^+ \rightarrow ClO_2 + H_2O \text{ (reduction)}$$

Multiply the reduction half-reaction by 2 to balance electrons.

$$ClO_{3-} + H_2O \rightarrow ClO_4^- + 2e^- + 2H^+ \text{ (oxidation)}$$
$$2ClO_3^- + 2e^- + 4H^+ \rightarrow 2ClO_2 + 2H_2O \text{ (reduction)}$$

Add the half-reactions and cancel like terms.

$$3ClO_3^- + 2H^+ \rightarrow ClO_4^- + 2ClO_2 + H_2O$$

Return one spectator H^+ ion to each side of the equation. On the left side, the H^+ ions join with the ClO_3^- ions. On the right side, the H^+ ion combines with the ClO_4^- ion.

$$3HClO_3(aq) \rightarrow HClO_4(aq) + 2ClO_2(g) + H_2O(l)$$

Chapter 21

Practice Problems

1. a. $Mg(s) + Pd^{2+}(aq) \rightarrow Mg^{2+}(aq) + Pd(s)$

$E^0_{cell} = +3.323$ V

$Mg|Mg^{2+}\|Pd^{2+}|Pd$

b. $Cd(s) + 2Cu^+(aq) \rightarrow Cd^{2+}(aq) + 2Cu(s)$

$E^0_{cell} = = +0.924$ V

$Cd|Cd^{2+}\|Cu^+|Cu$

c. $2Ce(s) + 6H^+(aq) \rightarrow 2Ce^{3+}(aq) + 3H_2(g)$

$E^0_{cell} = = +2.336$ V

$Ce|Ce^{3+}\|H^+|H_2$

2. a. $+1.247$ V   **b.** $+2.1201$ V

3. a. $+1.08$ V; spontaneous

b. -0.368 V; not spontaneous

c. -0.163 V; not spontaneous

d. $+0.736$ V; spontaneous

4. lead

5. primary battery

6. zinc

7. water

Chapter 21 Review

8. The most easily oxidized metals have the lowest (most negative) standard reduction potentials. The order from most easily oxidized to least easily oxidized is Ca, Mg, Zn, Pb, Pt.

9. a. cadmium

b. tin

c. cadmium

d. $Cd(s) + Sn^{2+}(aq) \rightarrow Cd^{2+}(aq) + Sn(s)$

e. $E^0_{cell} = E^0_{Sn^{2+}|Sn} - E^0_{Ca^{2+}|Ca}$

$\quad\quad = -0.1375 \text{ V} - (-0.4030 \text{ V})$

$\quad\quad = +0.2655 \text{ V}$

10. $E^0_{cell} = E^0_{reduction} - E^0_{oxidation}$

$\quad +2.197 \text{ V} = E^0_{I_2|I^-} - E^0_{oxidation}$

$\quad +2.197 \text{ V} = +0.5355 \text{ V} - E^0_{oxidation}$

$\quad E^0_{oxidation} = +0.5355 \text{ V} - 2.197 \text{ V}$

$\quad\quad\quad\quad = -1.662 \text{ V}$

From Table 21-1, the reduction half-reaction with this voltage is $Al^{3+}(aq) + 3e^- \rightarrow Al(s)$. Therefore, the probable oxidation half-reaction is $Al(s) \rightarrow Al^{3+}(aq) + 3e^-$.

11. Three methods used to minimize corrosion of iron may include applying a coat of paint, placing the iron in contact with a more easily oxidized metal, and galvanizing the iron by coating it with a layer of zinc.

12. A voltaic cell converts chemical energy to electrical energy by a spontaneous redox reaction. An electrolytic cell uses electrical energy to bring about a nonspontaneous redox reaction by the process of electrolysis.

Chapter 22

Practice Problems

1. a. 2-methylbutane

b. 3-ethyl-2,5-dimethylhexane

c. 4,7-diethyldecane

d. 3,5-diethyl-2,5-dimethyloctane

2. a. $CH_3CH_2CHCH_2CH_3$
$\quad\quad\quad\quad |$
$\quad\quad\quad CH_2CH_3$

b.

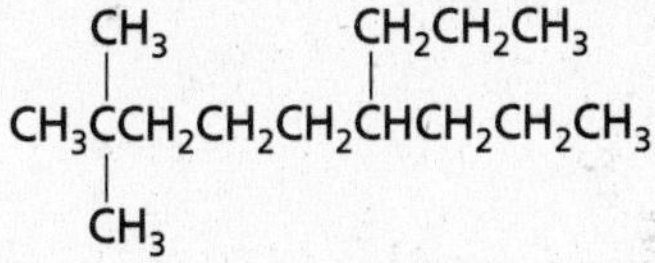

$$CH_3CCH_2CH_2CH_2CHCH_2CH_2CH_3$$

with CH_3 above (as CH_3C), CH_3 below the second carbon, and $CH_2CH_2CH_3$ above the sixth carbon.

c.

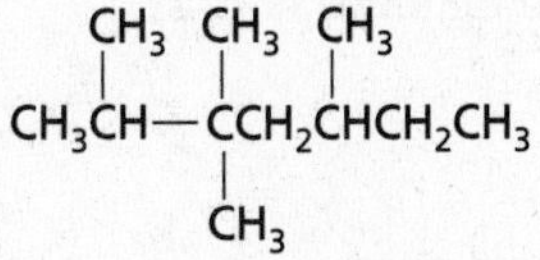

$$CH_3CH-CCH_2CHCH_2CH_3$$

with CH_3, CH_3, CH_3 above, and CH_3 below.

d.

$$CH_3CH_2CH-C-CHCH_2CH_2CH_3$$

with CH_3, CH_3, CH_3 above, and CH_2CH_3 below.

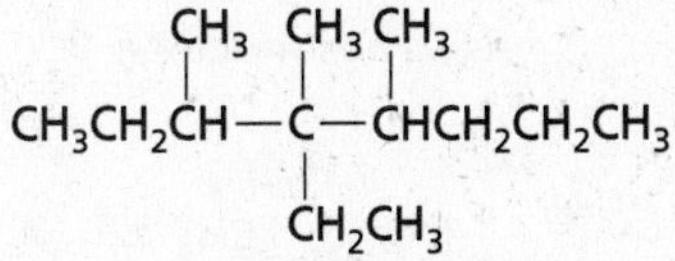

3. a. 1,2-diethylcyclopropane

b. 1,2,4-triethyl-7-methylcyclooctane

4. a. (cyclobutane with $CH_2CH_2CH_3$ substituent)

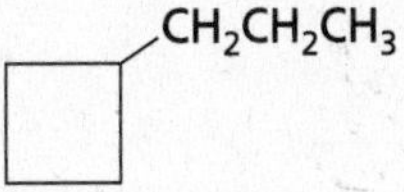

b. (cyclopentane with CH_2CH_3, CH_3, and CH_3 substituents)

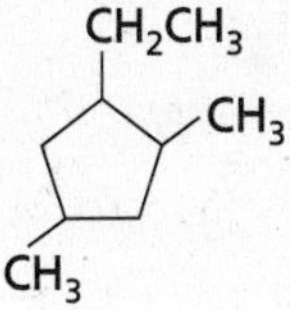

c. (cyclohexane with CH_3, CH_3, CH_3, CH_2CH_3, and CH_3CHCH_3 substituents)

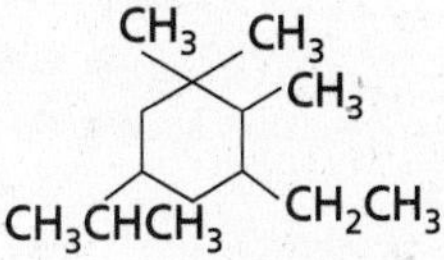

5. a. 3,5,5-trimethyl-3-heptene

b. 2,4-diethyl-1-methylcyclopentene

6. a.

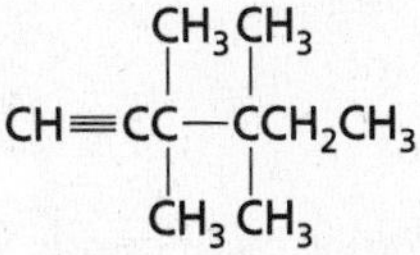

$$CH\!\equiv\!CC-CCH_2CH_3$$

with CH_3, CH_3 above and CH_3, CH_3 below.

b.

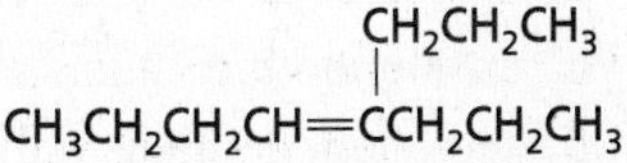

$$CH_3CH_2CH_2CH\!=\!CCH_2CH_2CH_3$$

with $CH_2CH_2CH_3$ above.

c.

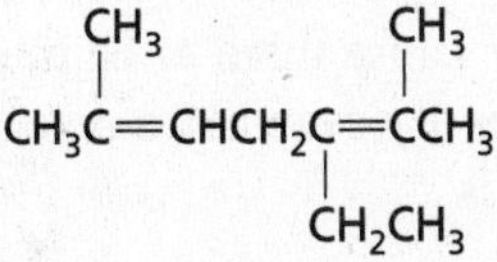

$$CH_3C\!=\!CHCH_2C\!=\!CCH_3$$

with CH_3, CH_3 above and CH_2CH_3 below.

7. a. isomers

   b. same compound

   c. same compound

8. 3-ethyl-2-hexene; its molecular formula is C_8H_{16}, whereas the molecular formula of the other compounds is C_7H_{14}.

9.

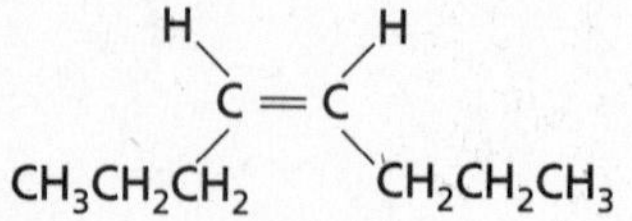

cis-4-octene 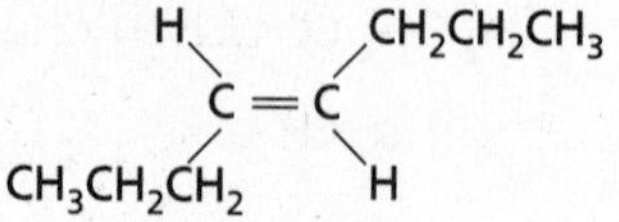   trans-4-octene

10. a.

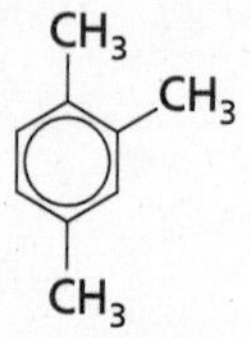

C_9H_{12}

b.

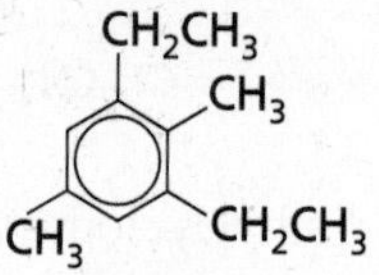

$C_{12}H_{18}$

Chapter 22 Review

11. The name is incorrect because the longest carbon chain contains eight carbon atoms. The correct numbering is as follows.

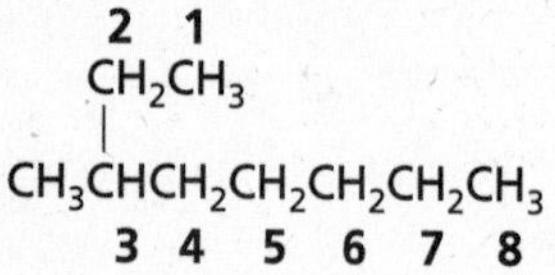

The correct name is 3-methyloctane.

12. a. saturated; contains only single bonds

   b. unsaturated; contains two double bonds

   c. unsaturated; contains a triple bond

   d. saturated; contains only single bonds

13. Alkanes, alkenes, and alkynes are nonpolar compounds with relatively low melting and boiling points and low solubilities in water. The double bonds of alkenes cause them to be more reactive than alkanes. Alkynes are generally even more reactive than alkenes because of their triple bonds.

14. Yes; both compounds have the molecular formula C_5H_8, but they have different carbon chains.

15. The compound is *trans*-2-hexene.

16. The aromatic compounds are 1-methyl-4-propylbenzene and anthracene because these compounds contain benzene rings as part of their molecular structures.

Chapter 23

Practice Problems

1. a. 2-bromo-1,1-dichloro-3-iodopropane

   b. 1-chloro-2,3-difluorobenzene

2. a. $CH_3CH_3 + Br_2 \rightarrow CH_3CH_2Br + HBr$

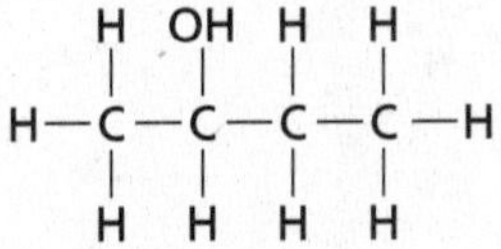

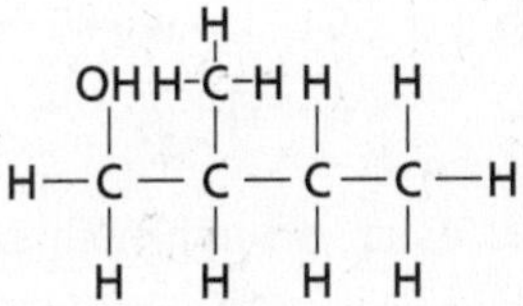

   b.

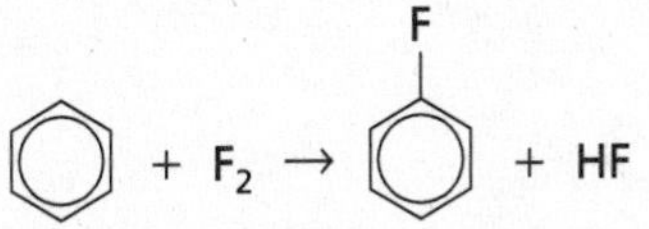

3. a. 2,3-hexanediol

   b. 2-methyl-2-propanol

4.

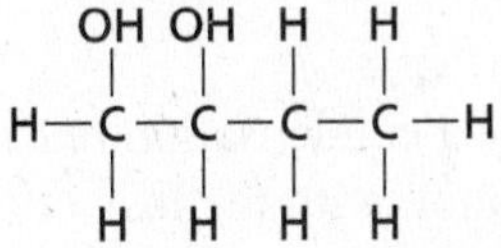

5. a. amine; 1,3-pentanediamine

   b. ether; methyl ether

6.

a.

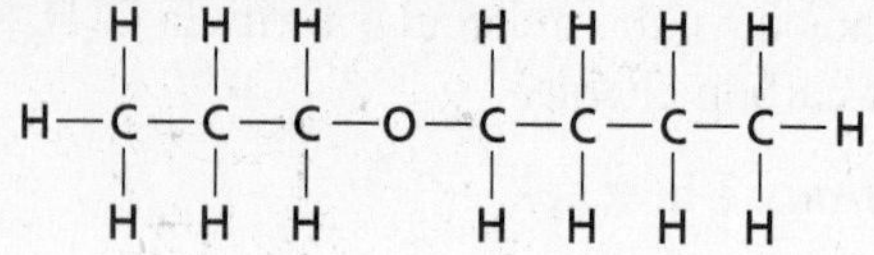

b.

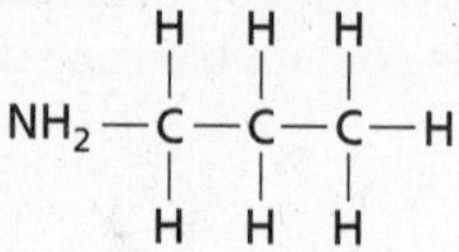

c.

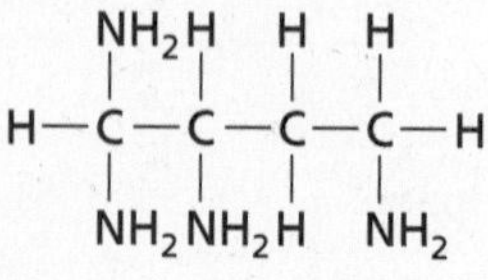

7. a. ketone; 2-pentanone
   b. aldehyde; propanal

8.

a.

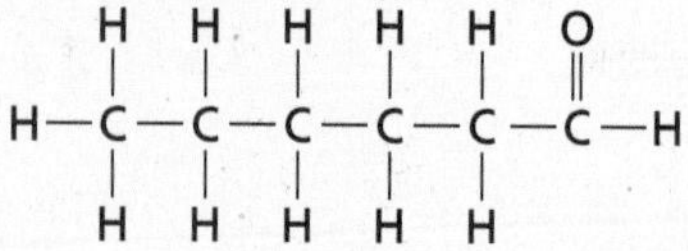

b.

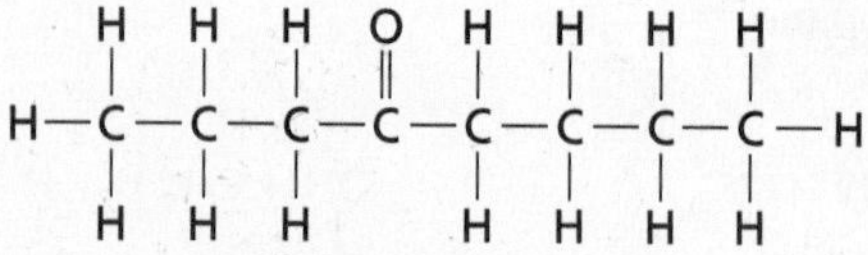

9. a. ester; propyl methanoate
   b. carboxylic acid; hexanoic acid
   c. amide; butanamide

10. Propyl methanoate can be made by a condensation reaction
    between methanoic acid and propanol, with the loss of a water
    molecule.

11.

a. 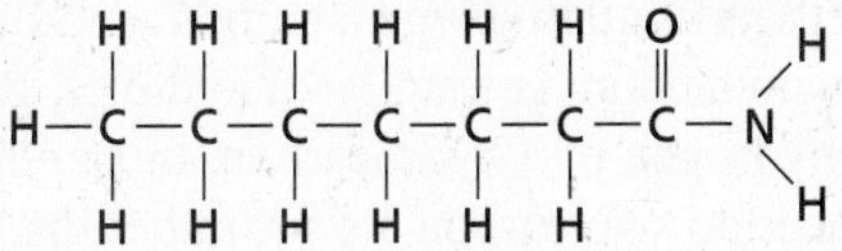 (structure: chain of carbons with hydrogens ending in amide)

H—C—C—C—C—C—C—C—N with H's

c. H—C—C—OH (with H and double-bond O)

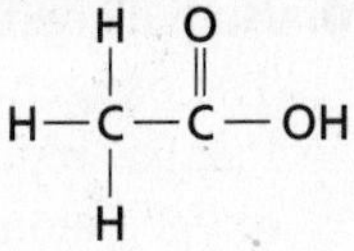

b. 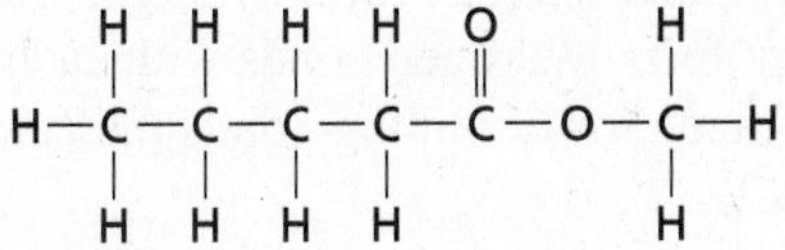 H—C—C—C—C—C—O—C—H (with H's and double-bond O)

12. a. elimination, dehydration

   b. addition, none of these

   c. elimination, dehydrogenation

13. a. butane

   b. 1,2-dibromoethane

   c. ethene and water

14. addition polymerization

15. condensation polymerization

16. thermoplastic polymer

17. thermosetting polymer

Chapter 23 Review

18. Halocarbons are named by use of the prefixes *fluoro-*, *chloro-*, *bromo-*, and *iodo-* plus the alkane name. If there is more than one kind of halogen, the atoms are listed alphabetically. Numbers are used to identify the positions of the halogens. The lowest possible position number is given to the substituent that comes first alphabetically.

19. A substitution reaction is one in which one atom or group of atoms in a molecule is replaced by another atom or group of atoms. An example is the formation of chloroethane and hydrogen chloride from ethane and chlorine. A condensation reaction is one in which two smaller organic molecules combine to form a more complex molecule, accompanied by the loss of a small molecule such as water. An example is the production of ethyl propanoate and water from ethanol and propanoic acid.

20. Both alcohols and ethers contain oxygen, carbon, and hydrogen. The oxygen atom in an alcohol is found in a hydroxyl group that has replaced a hydrogen of a hydrocarbon. In an ether, the oxygen atom is bonded to two carbon atoms, rather than to hydrogen. Alcohols are polar and can form hydrogen bonds. Ether molecules cannot form hydrogen bonds with each other and thus tend to have much lower boiling points than alcohols of similar size and mass.

21. A carbonyl group consists of an oxygen atom double-bonded to a carbon atom. In an aldehyde, the carbonyl group is located at the end of a carbon chain and is bonded to a carbon atom on one side and a hydrogen atom on the other. In a ketone, the carbon of the carbonyl group is bonded to two other carbon atoms.

22. A carboxylic acid is an organic compound that has a carboxyl group (–COOH). Carboxylic acids are polar, reactive, and acidic. They are named by changing the *-ane* suffix of the parent alkane to *-anoic acid*.

23. Esters can be made through a condensation reaction between a carboxylic acid and an alcohol. Amides can be made by replacing the –OH group of a carboxylic acid with a nitrogen atom bonded to other atoms.

24. In an elimination reaction, a combination of atoms is removed from two adjacent carbon atoms, forming an additional bond between the carbon atoms. Types of elimination reactions include dehydrogenation reactions and dehydration reactions. In an addition reaction, other atoms bond to each of two atoms bonded by double or triple covalent bonds. Types of addition reactions include hydrogenation reactions and hydration reactions.

25. A polymer is a large molecule consisting of many repeating structural units. Monomers are smaller molecules that bond to form a polymer. A structural unit is a repeating group of atoms formed by the bonding of the monomers.

26. In addition polymerization, all the atoms present in the monomers are retained in the polymer product. In condensation polymerization, monomers containing at least two functional groups combine with the loss of a small by-product, usually water.

Chapter 24

Practice Problems

1–4.

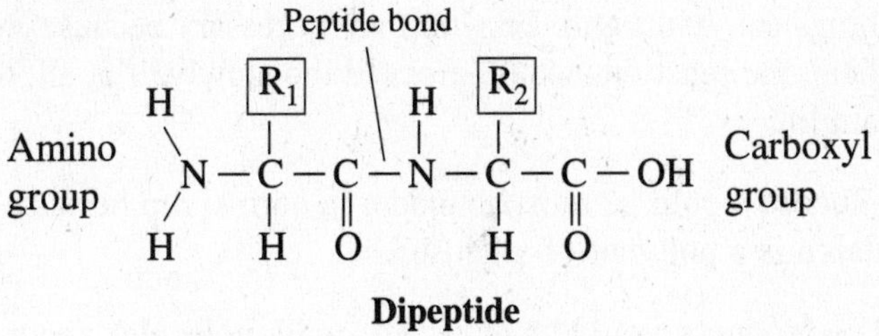

Peptide bond

Amino group

$$H-N-C-C-N-C-C-OH$$

Carboxyl group

Dipeptide

5. d	**7.** a	**9.** d	**11.** c
6. c	**8.** b	**10.** b	**12.** c

13. Carbon atoms in a saturated fatty acid are bonded only by single bonds. An unsaturated fatty acid contains at least one carbon-carbon double or triple bond.

14. Steroids do not consist of long chains, as do fatty acids and triglycerides.

15. b	**17.** c	**19.** e	**21.** f
16. g	**18.** a	**20.** d	

	Photosynthesis	Cellular Respiration	Fermentation
22. Releases energy		✓	✓
23. Uses energy	✓		
24. Produces glucose	✓		
25. Breaks down glucose		✓	✓
26. Produces oxygen	✓		
27. Uses oxygen		✓	
28. Catabolic process		✓	✓
29. Anabolic process	✓		

Chapter 24 Review

30. Enzymes speed up the chemical reactions in living things or allow the reactions to take place at a temperature that living things can withstand. Enzymes are necessary because, without them, the reactions would proceed too slowly, if at all, to sustain life.

31. Glucose would be most abundant in corn syrup because cornstarch is a polymer of glucose.

32. The products would be three fatty acid molecules and a molecule of glycerol.

33. DNA provides the information for the synthesis of proteins. Many of these proteins are enzymes that carry out life processes.

34. The purpose of cellular respiration is to release energy by breaking down carbohydrate molecules. Photosynthesis is the reverse of cellular respiration.

Chapter 25

Practice Problems

1. a. $^{208}_{82}Pb$

   b. $^{214}_{83}Bi$

   c. $^{234}_{90}Th$

   d. $^{210}_{82}Pb$

2. a. $^{3}_{2}He$

   b. $^{235}_{93}Np$

   c. $^{60}_{27}Co$

   d. $^{234}_{91}Pa$

3. a. 1:1

   b. 1:1

   c. 1.58:1

   d. 1.6:1

4. a. $^{210}_{84}Po \rightarrow ^{206}_{82}Pb + ^{4}_{2}He$

   b. $^{234}_{92}U \rightarrow ^{230}_{90}Th + ^{4}_{2}He$

   c. $^{222}_{86}Rn \rightarrow ^{218}_{84}Po + ^{4}_{2}He$

   d. $^{230}_{90}Th \rightarrow ^{226}_{88}Ra + ^{4}_{2}He$

5. a. $^{3}_{1}H \rightarrow ^{3}_{2}He + ^{0}_{-1}\beta$

   b. $^{235}_{92}U \rightarrow ^{235}_{93}Np + ^{0}_{-1}\beta$

   c. $^{60}_{26}Fe \rightarrow ^{60}_{27}Co + ^{0}_{-1}\beta$

   d. $^{234}_{90}Th \rightarrow ^{234}_{91}Pa + ^{0}_{-1}\beta$

6. a. $^{38}_{20}Ca$

   b. $^{142}_{60}Nd$

7. a. $^{256}_{101}\text{Md}$

b. $^{12}_{6}\text{C}$

8. a. 1.953 mg

b. 6.3 g

c. 10.0 g; 2.5 g

9. $3\,^{4}_{2}\text{He} \rightarrow\,^{12}_{6}\text{C}$

10. Sample answer:
$$^{180}_{72}\text{Hf} +\,^{95}_{41}\text{Nb} \rightarrow\,^{275}_{113}\text{Uut}$$

Chapter 25 Review

11. Radioactivity is the process by which materials emit radiation.

12. Neutrons add an attractive force to the nucleus because they have no charge but are subject to the strong nuclear force.

13. For atoms with low atomic number (<20), nuclei with a neutron-to-proton ratio of 1:1 tend to be the most stable. Neutron-to-proton ratio increases to about 1.5:1 for the largest stable atoms.

14. a. above

b. within

c. below

d. above

15. increase: positron emission, electron capture; decrease: beta decay

16. No; carbon dating is used to date the remains of organisms that are less than 24 000 years old.

17. Induced transmutation is the process of striking nuclei with high-velocity charged particles. Induced transmutation produces transuranium elements (atomic number ≥ 93).

18. Geiger counters, scintillation counters, and film badges are used to detect and measure radiation.

19. Nuclear fusion is the combining of nuclei to form a single nucleus. Nuclear fission is the splitting of a nucleus into fragments.

20. Radiation can damage living cells. Any exposure to radiation could possibly be harmful. A dose exceeding 500 rem can be fatal.

Chapter 26

Practice Problems

1. **a.** troposphere   **c.** exosphere
   b. stratosphere   **d.** troposphere

2. **a.** N_2 + high-energy UV $\rightarrow$ 2N
   b. O + high-energy UV $\rightarrow$ O^+ + e^-
   c. N_2 + high-energy UV $\rightarrow$ $N_2{}^+$ + e^-

3. **a.** condensation   **c.** precipitation
   b. evaporation

4. **a.** sterilization   **d.** sedimentation
   b. coarse filtration   **e.** aeration
   c. sand filtration

5. **a.** liquid   **c.** solid
   b. gaseous

6. **a.** More CO_2 in the atmosphere, due to increased exhalation of carbon dioxide, increased burning of fossil fuels by people, and elimination of trees to accommodate the people
   b. More CO_2 in the atmosphere, due to fewer plants taking in CO_2 for photosynthesis

7. Nitrogen fixation provides a mechanism for converting atmospheric nitrogen, which organisms cannot use, to nitrates, which plants can use to make complex nitrogen-containing compounds. Animals can then eat these plants to obtain nitrogen in a useable form.

Chapter 26 Review

8. Photochemical smog forms when sunlight reacts with pollutants in the air. Acid rain forms when sulfur- or nitrogen-containing pollutants in the air combine with moisture.

9. troposphere, stratosphere, mesosphere, thermosphere, exosphere

10. During photodissociation, high-energy UV solar radiation is absorbed by a molecule, causing its chemical bonds to break. During photoionization, a molecule or an atom absorbs enough

high-energy UV solar radiation to lose an electron, forming a positively charged particle.

11. CFCs photodissociate, freeing Cl atoms, which then react with ozone molecules. The Cl atoms act as a catalyst, converting O_3 to O_2.

12. Desalination is the removal of salts from seawater to make it usable for living things. Reverse osmosis and distillation are methods used to accomplish desalination.

13. The greenhouse effect is the natural warming of Earth's surface that occurs when certain gases in the atmosphere absorb some of the solar energy that is converted to heat and reflected from Earth's surface. Global warming is an increase in global temperatures due to increases in the greenhouse effect.

14. In the atmosphere, lightning and rain act together to convert N_2 into nitrate ions. Certain bacteria living in the soil and on the roots of some plants convert N_2 into nitrate ions.

15. A forest fire would increase the amount of CO_2 in the atmosphere by decreasing the number of plants taking in CO_2 for photosynthesis. Also, the burning of organic matter releases CO_2.

SOLVING PROBLEMS: A CHEMISTRY HANDBOOK

Appendices

Table A-1

SI Prefixes		
Prefix	**Symbol**	**Scientific notation**
femto	f	10^{-15}
pico	p	10^{-12}
nano	n	10^{-9}
micro	μ	10^{-6}
milli	m	10^{-3}
centi	c	10^{-2}
deci	d	10^{-1}
deka	da	10^{1}
hecto	h	10^{2}
kilo	k	10^{3}
mega	M	10^{6}
giga	G	10^{9}
tera	T	10^{12}
peta	P	10^{15}

Table A-2

Physical Constants		
Quantity	**Symbol**	**Value**
Atomic mass unit	amu	1.67×10^{-27} kg
Avogadro's number	N	6.02×10^{23} particles/mole
Ideal gas constant	R	8.314 L·kPa/mol·K 0.0821 L·atm/mol·K 62.4 mm Hg·L/mol·K 62.4 torr·L/mol·K
Mass of an electron	m_e	9.11×10^{-31} kg 5.49×10^{-4} amu
Mass of a neutron	m_n	1.675×10^{-27} kg 1.008 665 amu
Mass of a proton	m_p	1.673×10^{-27} kg 1.007 276 amu
Molar volume of ideal gas at STP	V	22.4 L/mol
Normal boiling point of water	T_b	373.15 K 100.0°C
Normal freezing point of water	T_f	273.15 K 0.00°C
Planck's constant	h	6.626×10^{-34} J·s
Speed of light in a vacuum	c	$2.997\ 925 \times 10^{8}$ m/s

Table A-3

Names and Charges of Polyatomic Ions	
1–	**2–**
Acetate, CH_3COO^-	Carbonate, CO_3^{2-}
Amide, NH_2^-	Chromate, CrO_4^{2-}
Astatate, AtO_3^-	Dichromate, $Cr_2O_7^{2-}$
Azide, N_3^-	Hexachloroplatinate,
Benzoate, $C_6H_5COO^-$	$\quad PtCl_6^{2-}$
Bismuthate, BiO_3^-	Hexafluorosilicate, SiF_6^{2-}
Bromate, BrO_3^-	Molybdate, MoO_4^{2-}
Chlorate, ClO_3^-	Oxalate, $C_2O_4^{2-}$
Chlorite, ClO_2^-	Peroxide, O_2^{2-}
Cyanide, CN^-	Peroxydisulfate, $S_2O_8^{2-}$
Formate, $HCOO^-$	Ruthenate, RuO_4^{2-}
Hydroxide, OH^-	Selenate, SeO_4^{2-}
Hypobromite, BrO^-	Selenite, SeO_3^{2-}
Hypochlorite, ClO^-	Silicate, SiO_3^{2-}
Hypophosphite, $H_2PO_2^-$	Sulfate, SO_4^{2-}
Iodate, IO_3^-	Sulfite, SO_3^{2-}
Nitrate, NO_3^-	Tartrate, $C_4H_4O_6^{2-}$
Nitrite, NO_2^-	Tellurate, TeO_4^{2-}
Perbromate, BrO_4^-	Tellurite, TeO_3^{2-}
Perchlorate, ClO_4^-	Tetraborate, $B_4O_7^{2-}$
Periodate, IO_4^-	Thiosulfate, $S_2O_3^{2-}$
Permanganate, MnO_4^-	Tungstate, WO_4^{2-}
Perrhenate, ReO_4^-	
Thiocyanate, SCN^-	
Vanadate, VO_3^-	

Table A-3, *continued*

Names and Charges of Polyatomic Ions, *continued*	
3−	**4−**
Arsenate, AsO_4^{3-}	Hexacyanoferrate(II),
Arsenite, AsO_3^{3-}	$\quad Fe(CN)_6^{4-}$
Borate, BO_3^{3-}	Orthosilicate, SiO_4^{4-}
Citrate, $C_6H_5O_7^{3-}$	Diphosphate, $P_2O_7^{4-}$
Hexacyanoferrate(III),	
$\quad Fe(CN)_6^{3-}$	
Phosphate, PO_4^{3-}	
Phosphite, PO_3^{3-}	
1+	**2+**
Ammonium, NH_4^+	Mercury(I), Hg_2^{2+}
Neptunyl(V), NpO_2^+	Neptunyl(VI), NpO_2^{2+}
Plutonyl(V), PuO_2^+	Plutonyl(VI), PuO_2^{2+}
Uranyl(V), UO_2^+	Uranyl(VI), UO_2^{2+}
Vanadyl(V), VO_2^+	Vanadyl(IV), VO^{2+}

Table A-4

Ionization Constants					
Substance	**Ionization Constant**	**Substance**	**Ionization Constant**	**Substance**	**Ionization Constant**
HCOOH	1.77×10^{-4}	HBO_3^{-2}	1.58×10^{-14}	HS^-	1.00×10^{-19}
CH_3COOH	1.75×10^{-5}	H_2CO_3	4.5×10^{-7}	HSO_4^-	1.02×10^{-2}
$CH_2ClCOOH$	1.36×10^{-3}	HCO_3^-	4.68×10^{-11}	H_2SO_3	1.29×10^{-2}
$CHCl_2COOH$	4.47×10^{-2}	HCN	6.17×10^{-10}	HSO_3^-	6.17×10^{-8}
CCl_3COOH	3.02×10^{-1}	HF	6.3×10^{-4}	$HSeO_4^-$	2.19×10^{-2}
HOOCCOOH	5.36×10^{-2}	HNO_2	5.62×10^{-4}	H_2SeO_3	2.29×10^{-3}
$HOOCCOO^-$	1.55×10^{-4}	H_3PO_4	7.08×10^{-3}	$HSeO_3^-$	4.79×10^{-9}
CH_3CH_2COOH	1.34×10^{-5}	$H_2PO_4^-$	6.31×10^{-8}	HBrO	2.51×10^{-9}
C_6H_5COOH	6.25×10^{-5}	HPO_4^{2-}	4.17×10^{-13}	HClO	2.9×10^{-8}
H_3AsO_4	6.03×10^{-3}	H_3PO_3	5.01×10^{-2}	HIO	3.16×10^{-11}
$H_2AsO_4^-$	1.05×10^{-7}	$H_2PO_3^-$	2.00×10^{-7}	NH_3	5.62×10^{-10}
H_3BO_3	5.75×10^{-10}	H_3PO_2	5.89×10^{-2}	H_2NNH_2	7.94×10^{-9}
$H_2BO_3^-$	1.82×10^{-13}	H_2S	9.1×10^{-8}	H_2NOH	1.15×10^{-6}

Table A-5

Electronegativities

Metal
Metalloid
Nonmetal

Group 1	Group 2												Group 13	Group 14	Group 15	Group 16	Group 17
1 **H** 2.20																	
3 **Li** 0.98	4 **Be** 1.57												5 **B** 2.04	6 **C** 2.55	7 **N** 3.04	8 **O** 3.44	9 **F** 3.98
11 **Na** 0.93	12 **Mg** 1.31												13 **Al** 1.61	14 **Si** 1.90	15 **P** 2.19	16 **S** 2.58	17 **Cl** 3.16
19 **K** 0.82	20 **Ca** 1.00	21 **Sc** 1.36	22 **Ti** 1.54	23 **V** 1.63	24 **Cr** 1.66	25 **Mn** 1.55	26 **Fe** 1.83	27 **Co** 1.88	28 **Ni** 1.91	29 **Cu** 1.90	30 **Zn** 1.65		31 **Ga** 1.81	32 **Ge** 2.01	33 **As** 2.18	34 **Se** 2.55	35 **Br** 2.96
37 **Rb** 0.82	38 **Sr** 0.95	39 **Y** 1.22	40 **Zr** 1.33	41 **Nb** 1.6	42 **Mo** 2.16	43 **Tc** 2.10	44 **Ru** 2.2	45 **Rh** 2.28	46 **Pd** 2.20	47 **Ag** 1.93	48 **Cd** 1.69		49 **In** 1.78	50 **Sn** 1.96	51 **Sb** 2.05	52 **Te** 2.1	53 **I** 2.66
55 **Cs** 0.79	56 **Ba** 0.89	57 **La** 1.10	72 **Hf** 1.3	73 **Ta** 1.5	74 **W** 1.7	75 **Re** 1.9	76 **Os** 2.2	77 **Ir** 2.2	78 **Pt** 2.2	79 **Au** 2.4	80 **Hg** 1.9		81 **Tl** 1.8	82 **Pb** 1.8	83 **Bi** 1.9	84 **Po** 2.0	85 **At** 2.2
87 **Fr** 0.7	88 **Ra** 0.9	89 **Ac** 1.1															

Elements not included on this table have no measured electronegativity.

Lanthanide series

58 **Ce** 1.12	59 **Pr** 1.13	60 **Nd** 1.14	61 **Pm** —	62 **Sm** 1.17	63 **Eu** —	64 **Gd** 1.20	65 **Tb** —	66 **Dy** 1.22	67 **Ho** 1.23	68 **Er** 1.24	69 **Tm** 1.25	70 **Yb** —	71 **Lu** 1.0

Actinide series

90 **Th** 1.3	91 **Pa** 1.5	92 **U** 1.7	93 **Np** 1.3	94 **Pu** 1.3

Table A-6

Specific Heat Values (J/g·°C)					
Substance	**c**	**Substance**	**c**	**Substance**	**c**
AlF_3	0.8948	Fe_3C	0.5898	$NaVO_3$	1.540
$BaTiO_3$	0.79418	$FeWO_4$	0.37735	$Ni(CO)_4$	1.198
BeO	1.020	HI	0.22795	PbI_2	0.1678
CaC_2	0.9785	K_2CO_3	0.82797	SF_6	0.6660
$CaSO_4$	0.7320	$MgCO_3$	0.8957	SiC	0.6699
CCl_4	0.85651	$Mg(OH)_2$	1.321	SiO_2	0.7395
CH_3OH	2.55	$MgSO_4$	0.8015	$SrCl_2$	0.4769
CH_2OHCH_2OH	2.413	MnS	0.5742	Tb_2O_3	0.3168
CH_3CH_2OH	2.4194	Na_2CO_3	1.0595	$TiCl_4$	0.76535
CdO	0.3382	NaF	1.116	Y_2O_3	0.45397
$CuSO_4 \cdot 5H_2O$	1.12				

Table A-7

Molal Freezing Point Depression and Boiling Point Elevation Constants				
Substance	K_{fp} **(C°/m)**	**Freezing Point (°C)**	K_{bp} **(C°/m)**	**Boiling Point (°C)**
Acetic acid	3.90	16.66	3.22	117.90
Benzene	5.12	5.533	2.53	80.100
Camphor	37.7	178.75	5.611	207.42
Cyclohexane	20.0	6.54	2.75	80.725
Cyclohexanol	39.3	25.15	—	—
Nitrobenzene	6.852	5.76	5.24	210.8
Phenol	7.40	40.90	3.60	181.839
Water	1.86	0.000	0.512	100.000

Table A-8

Heat of Formation Values			
ΔH_f° (kJ/mol) (concentration of aqueous solutions is 1M)			
Substance	ΔH_f°	**Substance**	ΔH_f°
$Ag(s)$	0	$H_3PO_3(aq)$	−964.4
$AgCl(s)$	−127.068	$H_3PO_4(aq)$	−1279.0
$AgCN(s)$	146.0	$H_2S(g)$	−20.63
Al_2O_3	−1675.7	$H_2SO_3(aq)$	−608.81
$BaCl_2(aq)$	−871.95	$H_2SO_4(aq)$	−814.0
$BaSO_4$	−1473.2	$HgCl_2(s)$	−224.3
$BeO(s)$	−609.6	$Hg_2Cl_2(s)$	−265.22
$BiCl_3(s)$	−379.1	$Hg_2SO_4(s)$	−743.12
$Bi_2S_3(s)$	−143.1	$I_2(s)$	0
Br_2	0	$KBr(s)$	−393.798
$CCl_4(l)$	−128.2	$KMnO_4(s)$	−837.2
$CH_4(g)$	−74.81	KOH	−424.764
$C_2H_2(g)$	226.73	$LiBr(s)$	−351.213
$C_2H_4(g)$	52.26	$LiOH(s)$	−484.93
$C_2H_6(g)$	−84.68	$Mn(s)$	0
$CO(g)$	−110.525	$MnCl_2(aq)$	−555.05
$CO_2(g)$	−393.509	$Mn(NO_3)_2(aq)$	−635.5
$CS_2(l)$	89.70	$MnO_2(s)$	−520.03
$Ca(s)$	0	$MnS(s)$	−214.2
$CaCO_3(s)$	−1206.9	$N_2(g)$	0
$CaO(s)$	−635.1	$NH_3(g)$	−46.11
$Ca(OH)_2(s)$	−986.09	$NH_4Br(s)$	−270.83
$Cl_2(g)$	0	$NO(g)$	90.25
$Co_3O_4(s)$	−891	$NO_2(g)$	33.18
$CoO(s)$	−237.94	$N_2O(g)$	82.05
$Cr_2O_3(s)$	−1139.7	$Na(s)$	0
$CsCl(s)$	−443.04	$NaBr(s)$	−361.062
$Cs_2SO_4(s)$	−1443.02	$NaCl(s)$	−411.153
$CuI(s)$	−67.8	$NaHCO_3(s)$	−950.8
$CuS(s)$	−53.1	$NaNO_3(aq)$	−447.48
$Cu_2S(s)$	−79.5	$NaOH(s)$	−425.609
$CuSO_4(s)$	−771.36	$Na_2CO_3(s)$	−1130.7
$F_2(g)$	0	$Na_2S(aq)$	−447.3
$FeCl_3(s)$	−399.49	$Na_2SO_4(s)$	−1387.08
$FeO(s)$	−272.0	$NH_4Cl(s)$	−314.4
$FeS(s)$	−100.0	$O_2(g)$	0
$Fe_2O_3(s)$	−824.2	$P_4O_6(s)$	−1640.1
$Fe_3O_4(s)$	−1118.4	$P_4O_{10}(s)$	−2984.0
$H(g)$	217.965	$PbBr_2(s)$	−278.7
$H_2(g)$	0	$PbCl_2(s)$	−359.41
$HBr(g)$	−36.40	$SF_6(g)$	−1220.5
$HCl(g)$	−92.307	$SO_2(g)$	−296.830
$HCl(aq)$	−167.159	$SO_3(g)$	−454.51
$HCN(aq)$	108.9	$SrO(s)$	−592.0
$HCHO$	−108.57	$TiO_3(s)$	−939.7
$HCOOH(l)$	−424.72	$TlI(s)$	−123.5
$HF(g)$	−271.1	$UCl_4(s)$	−1019.2
$HI(g)$	26.48	$UCl_5(s)$	−1059
$H_2O(l)$	−285.830	$ZnCl_2(aq)$	−488.19
$H_2O(g)$	−241.818	$ZnO(s)$	−348.28
$H_2O_2(l)$	−187.8	$ZnSO_4(aq)$	−1063.15

Table A-9

PERIODIC TABLE OF THE ELEMENTS

Element — Hydrogen
Atomic number — 1 — State of matter
Symbol — H
Atomic mass — 1.008

- Gas
- Liquid
- Solid
- Synthetic

1A 1	2A 2		3B 3	4B 4	5B 5	6B 6	7B 7	8B 8	9
Hydrogen 1 H 1.008									
Lithium 3 Li 6.941	**Beryllium** 4 Be 9.012								
Sodium 11 Na 22.990	**Magnesium** 12 Mg 24.305								
Potassium 19 K 39.098	**Calcium** 20 Ca 40.078		**Scandium** 21 Sc 44.956	**Titanium** 22 Ti 47.867	**Vanadium** 23 V 50.942	**Chromium** 24 Cr 51.996	**Manganese** 25 Mn 54.938	**Iron** 26 Fe 55.845	**Cobalt** 27 Co 58.933
Rubidium 37 Rb 85.468	**Strontium** 38 Sr 87.62		**Yttrium** 39 Y 88.906	**Zirconium** 40 Zr 91.224	**Niobium** 41 Nb 92.906	**Molybdenum** 42 Mo 95.94	**Technetium** 43 Tc (98)	**Ruthenium** 44 Ru 101.07	**Rhodium** 45 Rh 102.906
Cesium 55 Cs 132.905	**Barium** 56 Ba 137.327		**Lanthanum** 57 La 138.906	**Hafnium** 72 Hf 178.49	**Tantalum** 73 Ta 180.948	**Tungsten** 74 W 183.84	**Rhenium** 75 Re 186.207	**Osmium** 76 Os 190.23	**Iridium** 77 Ir 192.217
Francium 87 Fr (223)	**Radium** 88 Ra (226)		**Actinium** 89 Ac (227)	**Rutherfordium** 104 Rf (261)	**Dubnium** 105 Db (262)	**Seaborgium** 106 Sg (266)	**Bohrium** 107 Bh (264)	**Hassium** 108 Hs (277)	**Meitnerium** 109 Mt (268)

The number in parentheses is the mass number of the longest lived isotope for that element.

Lanthanide series	**Cerium** 58 Ce 140.116	**Praseodymium** 59 Pr 140.908	**Neodymium** 60 Nd 144.24	**Promethium** 61 Pm (145)	**Samarium** 62 Sm 150.36	**Europium** 63 Eu 151.964
Actinide series	**Thorium** 90 Th 232.038	**Protactinium** 91 Pa 231.036	**Uranium** 92 U 238.029	**Neptunium** 93 Np (237)	**Plutonium** 94 Pu (244)	**Americium** 95 Am (243)

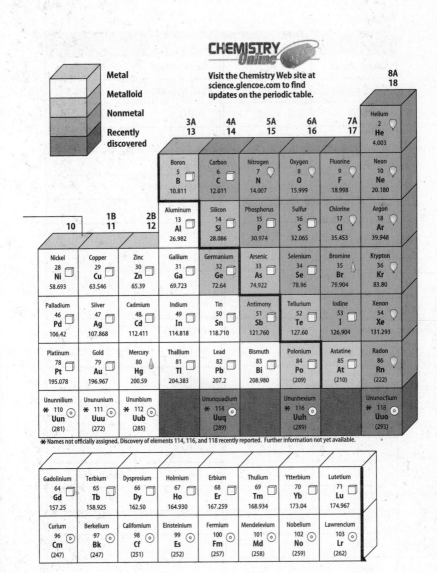

Metal

Metalloid

Nonmetal

Recently discovered

CHEMISTRY
Online

Visit the Chemistry Web site at
science.glencoe.com to find
updates on the periodic table.

	1B 11	2B 12	3A 13	4A 14	5A 15	6A 16	7A 17	8A 18
10								Helium 2 He 4.003
			Boron 5 B 10.811	Carbon 6 C 12.011	Nitrogen 7 N 14.007	Oxygen 8 O 15.999	Fluorine 9 F 18.998	Neon 10 Ne 20.180
			Aluminum 13 Al 26.982	Silicon 14 Si 28.086	Phosphorus 15 P 30.974	Sulfur 16 S 32.065	Chlorine 17 Cl 35.453	Argon 18 Ar 39.948
Nickel 28 Ni 58.693	Copper 29 Cu 63.546	Zinc 30 Zn 65.39	Gallium 31 Ga 69.723	Germanium 32 Ge 72.64	Arsenic 33 As 74.922	Selenium 34 Se 78.96	Bromine 35 Br 79.904	Krypton 36 Kr 83.80
Palladium 46 Pd 106.42	Silver 47 Ag 107.868	Cadmium 48 Cd 112.411	Indium 49 In 114.818	Tin 50 Sn 118.710	Antimony 51 Sb 121.760	Tellurium 52 Te 127.60	Iodine 53 I 126.904	Xenon 54 Xe 131.293
Platinum 78 Pt 195.078	Gold 79 Au 196.967	Mercury 80 Hg 200.59	Thallium 81 Tl 204.383	Lead 82 Pb 207.2	Bismuth 83 Bi 208.980	Polonium 84 Po (209)	Astatine 85 At (210)	Radon 86 Rn (222)
Ununnilium * 110 Uun (281)	Unununium * 111 Uuu (272)	Ununbium * 112 Uub (285)	Ununquadium * 114 Uuq (289)		Ununhexium * 116 Uuh (289)			Ununoctium * 118 Uuo (293)

* Names not officially assigned. Discovery of elements 114, 116, and 118 recently reported. Further information not yet available.

Gadolinium 64 Gd 157.25	Terbium 65 Tb 158.925	Dysprosium 66 Dy 162.50	Holmium 67 Ho 164.930	Erbium 68 Er 167.259	Thulium 69 Tm 168.934	Ytterbium 70 Yb 173.04	Lutetium 71 Lu 174.967
Curium 96 Cm (247)	Berkelium 97 Bk (247)	Californium 98 Cf (251)	Einsteinium 99 Es (252)	Fermium 100 Fm (257)	Mendelevium 101 Md (258)	Nobelium 102 No (259)	Lawrencium 103 Lr (262)

Table A-10

Solubility Product Constants at 298 K			
Compound	K_{sp}	**Compound**	K_{sp}
Carbonates		**Phosphates**	
$BaCO_3$	2.6×10^{-9}	$AlPO_4$	9.8×10^{-21}
$CaCO_3$	3.4×10^{-9}	$Ca_3(PO_4)_2$	2.1×10^{-33}
$CuCO_3$	2.5×10^{-10}	$Mg_3(PO_4)_2$	1.0×10^{-24}
$PbCO_3$	7.4×10^{-14}	**Hydroxides**	
$MgCO_3$	6.8×10^{-6}	$Al(OH)_3$	4.6×10^{-33}
Ag_2CO_3	8.5×10^{-12}	$Ca(OH)_2$	5.0×10^{-6}
$ZnCO_3$	1.5×10^{-10}	$Cu(OH)_2$	2.2×10^{-20}
Hg_2CO_3	3.6×10^{-17}	$Fe(OH)_2$	4.9×10^{-17}
Chromates		$Fe(OH)_3$	2.8×10^{-39}
$BaCrO_4$	1.2×10^{-10}	$Mg(OH)_2$	5.6×10^{-12}
$PbCrO_4$	2.3×10^{-13}	$Zn(OH)_2$	3×10^{-17}
Ag_2CrO_4	1.1×10^{-12}	**Sulfates**	
Halides		$BaSO_4$	1.1×10^{-10}
CaF_2	3.5×10^{-11}	$CaSO_4$	4.9×10^{-5}
$PbBr_2$	6.6×10^{-6}	$PbSO_4$	2.5×10^{-8}
$PbCl_2$	1.7×10^{-5}	Ag_2SO_4	1.2×10^{-5}
PbF_2	3.3×10^{-8}		
PbI_2	9.8×10^{-9}		
$AgCl$	1.8×10^{-10}		
$AgBr$	5.4×10^{-13}		
AgI	8.5×10^{-17}		

Table A-11

Standard Reduction Potentials at 25°C, 1 atm, and 1M Ion Concentration	
Half-reaction	**E^0 (V)**
$Li^+ + e^- \rightarrow Li$	-3.0401
$Cs^+ + e^- \rightarrow Cs$	-3.026
$K^+ + e^- \rightarrow K$	-2.931
$Ba^{2+} + 2e^- \rightarrow Ba$	-2.912
$Ca^{2+} + 2e^- \rightarrow Ca$	-2.868
$Na^+ + e^- \rightarrow Na$	-2.71
$Mg^{2+} + 2e^- \rightarrow Mg$	-2.372
$Ce^{3+} + 3e^- \rightarrow Ce$	-2.336
$H_2 + 2e^- \rightarrow 2H^-$	-2.323
$Nd^{3+} + 3e^- \rightarrow Nd$	-2.1
$Be^{2+} + 2e^- \rightarrow Be$	-1.847
$Al^{3+} + 3e^- \rightarrow Al$	-1.662
$Mn^{2+} + 2e^- \rightarrow Mn$	-1.185
$Cr^{2+} + 2e^- \rightarrow Cr$	-0.913
$2H_2O + 2e^- \rightarrow H_2 + 2OH^-$	-0.8277
$Zn^{2+} + 2e^- \rightarrow Zn$	-0.7618
$Cr^{3+} + 3e^- \rightarrow Cr$	-0.744
$Ga^{3+} + 3e^- \rightarrow Ga$	-0.549
$2CO_2 + 2H^+ + 2e^- \rightarrow H_2C_2O_4$	-0.49
$S + 2e^- \rightarrow S^{2-}$	-0.47627
$Fe^{2+} + 2e^- \rightarrow Fe$	-0.447
$Cr^{3+} + e^- \rightarrow Cr^{2+}$	-0.407
$Cd^{2+} + 2e^- \rightarrow Cd$	-0.4030
$PbI_2 + 2e^- \rightarrow Pb + 2I^-$	-0.365
$PbSO_4 + 2e^- \rightarrow Pb + SO_4^{2-}$	-0.3588
$Co^{2+} + 2e^- \rightarrow Co$	-0.28
$Ni^{2+} + 2e^- \rightarrow Ni$	-0.257
$Sn^{2+} + 2e^- \rightarrow Sn$	-0.1375
$Pb^{2+} + 2e^- \rightarrow Pb$	-0.1262
$Fe^{3+} + 3e^- \rightarrow Fe$	-0.037
$2H^+ + 2e^- \rightarrow H_2$	0.0000

Table A-11, *continued*

Standard Reduction Potentials at 25°C, 1 atm, and 1*M* Ion Concentration, *continued*	
Half-reaction	**E^0 (V)**
$Sn^{4+} + 2e^- \rightarrow Sn^{2+}$	0.151
$Cu^{2+} + e^- \rightarrow Cu^+$	0.153
$SO_4^{2-} + 4H^+ + 2e^- \rightarrow H_2SO_3 + H_2O$	0.172
$Bi^{3+} + 3e^- \rightarrow Bi$	0.308
$Cu^{2+} + 2e^- \rightarrow Cu$	0.3419
$O_2 + 2H_2O + 4e^- \rightarrow 4OH^-$	0.401
$Cu^+ + e^- \rightarrow Cu$	0.521
$I_2 + 2e^- \rightarrow 2I^-$	0.5355
$O_2 + 2H^+ + 2e^- \rightarrow H_2O_2$	0.695
$Fe^{3+} + e^- \rightarrow Fe^{2+}$	0.771
$NO_3^- + 2H^+ + e^- \rightarrow NO_2 + H_2O$	0.775
$Hg_2^{2+} + 2e^- \rightarrow 2Hg$	0.7973
$Ag^+ + e^- \rightarrow Ag$	0.7996
$Hg^{2+} + 2e^- \rightarrow Hg$	0.851
$2Hg^{2+} + 2e^- \rightarrow Hg_2^{2+}$	0.920
$Pd^{2+} + 2e^- \rightarrow Pd$	0.951
$NO_3^- + 4H^+ + 3e^- \rightarrow NO + 2H_2O$	0.957
$Br_2(l) + 2e^- \rightarrow 2Br^-$	1.066
$Ir^{3+} + 3e^- \rightarrow Ir$	1.156
$Pt^{2+} + 2e^- \rightarrow Pt$	1.18
$O_2 + 4H^+ + 4e^- \rightarrow 2H_2O$	1.229
$Cl_2 + 2e^- \rightarrow 2Cl^-$	1.35827
$Au^{3+} + 2e^- \rightarrow Au^+$	1.401
$Au^{3+} + 3e^- \rightarrow Au$	1.498
$MnO_4^- + 8H^+ + 5e^- \rightarrow Mn^{2+} + 4H_2O$	1.507
$Au^+ + e^- \rightarrow Au$	1.692
$H_2O_2 + 2H^+ + 2e^- \rightarrow 2H_2O$	1.776
$Co^{3+} + e^- \rightarrow Co^{2+}$	1.92
$S_2O_8^{2-} + 2e^- \rightarrow 2SO_4^{2-}$	2.010
$O_3 + 2H^+ + 2e^- \rightarrow O_2 + H_2O$	2.076
$F_2 + 2e^- \rightarrow 2F^-$	2.866

Logarithms and Antilogarithms

When you perform calculations, such as using the half-life of carbon to determine the age of a prehistoric skull or the pH of household products, you may need to use the log or antilog function on your calculator. A logarithm (log) is the power or exponent to which a number, called a base, must be raised in order to obtain a given positive number. This textbook uses common logarithms based on a base of ten. Therefore, the common log of any number is the power to which ten is raised to equal that number. Examine **Table B-1.** Note the log of each number is the power of ten for the exponent of that number. For example, the common log of 100 is two and the common log of 0.01 is -2.

$$\log 10^2 = 2$$
$$\log 10^{-2} = -2$$

A common log can be written in the following general form.

If $10^n = y$, then $\log y = n$.

In each example in **Table B-1,** the log can be determined by inspection. How do you express the common log of 5.34×10^5? Because logarithms are exponents, they have the same properties as exponents. See **Table B-2.**

$$\log 5.34 \times 10^5 = \log 5.34 + \log 10^5$$

Table B-1

Comparison Between Exponents and Logs	
Exponent	**Logarithm**
$10^0 = 1$	$\log 1 = 0$
$10^1 = 10$	$\log 10 = 1$
$10^2 = 100$	$\log 100 = 2$
$10^{-1} = 0.1$	$\log 0.1 = -1$
$10^{-2} = 0.01$	$\log 0.01 = -2$

Most scientific calculators have a button labeled $\boxed{\text{log}}$ and, in most cases, the number is simply entered and the log button is pushed to display the log of the number. Note that there is the same number of digits after the decimal in the log as there are significant figures in the original number entered.

$$\log 5.34 \times 10^5 = \log 5.34 + \log 10^5 = 0.728 + 5 = 5.728$$

Suppose the pH of aqueous ammonia equals 9.54 and you are asked to find the concentration of the hydrogen ions in that solution. By definition, pH $= -\log [H^+]$. Compare this to the general equation for the common log.

Equation for pH: $\qquad pH = -\log [H^+]$

General equation: $\qquad y = \log 10^n$

To solve the equation for $[H^+]$, you must follow the reverse process and calculate the antilogarithm (antilog) of -9.54 to find $[H^+]$.

Antilogs are the reverse of logs. To find the antilog, use a scientific calculator to input the value of the log. Then, use the inverse function and press the log button.

If $n =$ antilog y, then $y = 10^n$.

Thus, $[H^+] =$ antilog$(-9.54) = 10^{-9.54} = 10^{0.46 +(-10)}$

$$= 10^{0.46} \times 10^{-10}$$

$$= 2.9 \times 10^{-10} M$$

Check the instruction manual for your calculator. The exact procedure to calculate logs and antilogs may vary.

Table B-2

Properties of Exponents	
Exponential Notation	**Logarithm**
$10^A \times 10^B = 10^{A + B}$	$\log (A \times B) = \log A + \log B$
$10^A \div 10^B = 10^{A - B}$	$\log (A \div B) = \log A - \log B$
A^B	$(\log A) \times B$

INDEX

A

Accuracy of measurement, 14
Acid-base indicator, 198
Acidic solution, 189
Acid ionization constant, 191, 197–198
Acids: amino, 245; versus bases, 189; DNA and RNA, 249–250; fatty, 247; monoprotic and polyprotic, 190; naming, 82–83; neutralizing, 198–200; nucleic, 249–250; strength of, 190–191; *See also* pH
Actinide series, 69
Activated complex, 171
Activation energy, 171
Activity series, 93–94
Actual yield, 122
Addition polymerization, 243
Addition reaction, 241
Adenosine diphosphate (ADP), 250
Adenosine triphosphate (ATP), 250
Alcohol, 235
Aldehyde, 237
Aliphatic compounds, 231
Alkali metals, 54, 63–64
Alkaline earth metals, 54, 64
Alkanes, 221–224, 226
Alkenes, 226–228
Alkyl halide, 233
Alkynes, 228
Allotropes, 66
Alloy, 26, 77
Alpha particles, 39
Alpha radiation, 39
Amide, 239
Amine, 235

Amino acid, 245
Amorphous solid, 132
Ampere, 7
Anabolism, 250
Analytical chemistry, 2
Anion, 71–72
Anode, 214
Applied research, 4
Aqueous solutions, 95–98
Aromatic compounds, 230–231
Arrhenius model, 189
Aryl halide, 233
Atmosphere, 128, 263–265
Atomic emission spectrum, 42–43
Atomic mass, 37–38
Atomic number, 33–34
Atomic radius, 57–58
Atoms: differences in, 33–39; early theories, 31–32; electron configurations, 45–51; light and quantized energy, 41–43; in makeup of matter, 2, 31; and quantum theory, 43–45; sub-atomic particles and nuclear atoms, 31–33; unstable nuclei and radioactive decay, 39–40; *See also* Matter
ATP (adenosine triphosphate), 250
Attractive forces, 130
Aufbau principle, 46
Avogadro's number, 99
Avogadro's principle, 140–141

B

Balanced chemical equation, 90–92
Band of stability, 255–256
Bar graph, 18

Barometer, 127–128
Base ionization constant, 192
Bases, 189, 191–192, 198–200;
   See also pH
Base unit, 7–8
Basic solution, 189
Batteries, 218–220; See also
   Electrochemistry; Voltaic cells
Becquerel, Henri, 253
Benzene, 230–231
Beta particles, 39
Beta radiation, 39
Biochemistry, 2
Bohr, Niels, 43
Bohr atomic model, 43
Boiling point, 133
Boiling point elevation, 156
Boltzmann, Ludwig, 125
Boron, 65
Boyle's law, 137–138
Brønsted-Lowry model, 189
Brownian motion, 158
Buffer, 200

C
Calorimeter, 160–161
Candela, 7
Capillary action, 131–132
Carbohydrates, 246–247
Carbon, 65–66
Carbon cycle, 268–269
Carbon dating, 259
Carbonyl compounds, 237–240
Carbonyl group, 237
Carboxyl group, 239
Catabolism, 250
Catalysts, 124, 172
Cathode, 214
Cathode ray, 31
Cation, 71–72
Cellular respiration, 251
Celsius scale, 10
CFCs (chlorofluorocarbons), 1,
   3, 4
Charles's law, 138–139

Chemical, defined, 1
Chemical bond, 71
Chemical change, 23; enthalpy,
   162, 163–166; heat in,
   159–162; reaction spontaneity,
   166–168; thermochemical
   equations, 162–163; See also
   Chemical reaction
Chemical equations, 23, 90–92
Chemical equilibrium, 177,
   181–183; See also Equilibrium
   constant
Chemical properties, 21–22
Chemical reaction, 23; in aque-
   ous solutions, 95–98;
   classifying, 92–95; equations
   representing, 89–92; See also
   Chemical change
Chemistry, scope of, 1, 2
Chlorofluorocarbons (CFCs), 1,
   3, 4
Circle graph, 18
Coefficients, 90
Colligative properties, 155–157
Collision theory, 171
Colloid, 158
Combined gas law, 139–140
Combustion reaction, 92
Common ion effect, 188
Complete ionic equation, 95
Complex reaction, 175–176
Compounds, 27–28; aromatic and
   aliphatic, 230–231; carbonyl,
   237–240; moles of, 103–104;
   naming, 83; See also Ionic
   compounds
Compression, 126, 131
Concentration, 150–155,
   171–172
Conclusion, defined, 3
Condensation, 134
Condensation polymerization,
   243
Condensation reaction, 239
Conjugate acid, 189

Solving Problems: A Chemistry Handbook

Conjugate acid-base pair, 189
Conjugate base, 189
Contact process, 124
Control, defined, 3
Conversion factor, 13
Coordinate covalent bond, 86
Corrosion, 219–220
Covalent bond, 79–82; electronegativity and polarity, 87–88; molecular shape, 86–87; molecular structures, 83–86; naming molecules, 82–83
Covalent network solid, 132
Critical mass, 260
Crystalline solid, 132, 133
Curie, Marie, 253
Curie, Pierre, 253
Cycloalkanes, 224–226

D
Dalton, John, 31, 39, 128
Dalton's atomic theory, 31
Dalton's law of partial pressure, 128–129
Data, graphing, 18–20
d-block elements, 69–70
de Broglie, Louis, 43, 44
de Broglie's waves, 43
Decomposition reaction, 92, 204
Dehydration reaction, 241
Dehydrogenation, 241
Delocalized electrons, 77
Democritus, 31
Density, 8–9, 126, 131
Deoxyribonucleic acid (DNA), 249–250
Dependent variable, 3
Deposition, 134
Derived unit, 8
Desalinization, 266
Diagonal relationship, 63
Diffusion, 126, 127
Dimensional analysis, 13–14
Dipole-dipole forces, 130

Dispersion forces, 130
DNA (deoxyribonucleic acid), 249–250
Double-replacement reaction, 94, 204
Dry cell, 218

E
Effusion, 126
Einstein, Albert, 42, 43, 260
Elastic collision, 125
Electric current, 7
Electrochemical cells, 213; *See also* Voltaic cells
Electrochemistry, 213, 218–220; *See also* Voltaic cells
Electrolysis, 220
Electrolytic cell, 220
Electromagnetic radiation, 41
Electron, 31–32, 34
Electron capture, 256
Electron configuration, 45–49, 49–51
Electron-dot structures, 49–51
Electronegativity, 60–61, 87–88
Electron sea model, 77
Elements: classification of, 55–57; d-block and f-block, 69–70; defined, 27; p-block, 65–69; s-block, 63–65; transuranium, 258; *See also* *specific elements*
Elimination reaction, 241
Empirical formula, 106–107
Endothermic reaction, 81
Energy: activation, 171; defined, 159; free, 167; ionization, 59; lattice, 73–74; levels, 44; in phase changes, 133–134
Energy sublevel, 44
Energy sublevel diagram, 47
Enthalpy, 162, 163–166
Enthalpy of combustion, 162
Enthalpy of reaction, 162–163
Entropy, 166–167

Solving Problems: A Chemistry Handbook

Hybridization, 86
Hydrates, 110–111
Hydration, 147
Hydration reaction, 241
Hydrocarbons: alkanes, 221–224, 226; alkenes, 226–228; alkynes, 228; aromatic, 230–231; cycloalkanes, 224–226; isomers, 229–230; *See also* Substituted hydrocarbons
Hydrogen, 63
Hydrogenation reaction, 241
Hydrogen bonds, 130
Hydrosphere, 266–267; *See also* Water
Hydroxyl group, 235
Hypothesis, 3

I

Ideal gas constant, 142–143, 142–144
Ideal gases, 137
Immiscible liquid, 147
Independent variable, 3
Induced transmutation, 257–258
Inhibitors, 172
Inner transition metals, 69
Inorganic chemistry, 2
Insoluble substance, 147
Intermediate, in reaction rate, 176
Intramolecular forces, 130
Ionic bonds, 72
Ionic compounds: and chemical bonds, 71–72; formation and nature of, 72–74; metallic bonds, 77–78; names and formulas for, 74–77; *See also* Compounds
Ionic radius, 58
Ionic solid, 132
Ionization energy, 59
Ionizing radiation, 261
Ion product constant for water, 192–193
Isomers, 229–230
Isotopes, 34–35

K

Kelvin, 7, 10
Kelvin scale, 10
Ketone, 238
Kilogram, 7
Kinetic-molecular theory, 125–126, 131

L

Laboratory safety, 4–5
Lanthanide series, 69
Lattice energy, 73–74
Law of chemical equilibrium, 177
Law of conservation of mass, 24–25, 113
Law of definite proportions, 28
Lead-acid batteries, 218
Le Châtelier's principle, 181
Length, measuring, 7
Lewis, G. N., 49
Lewis structure, 79, 84–85
Light, 7, 41–43
Limiting reactant, 120–122
Line graph, 18
Lipids, 247–248
Liquids, 22, 131–132
Liter, 8
Lithium batteries, 219
Lithosphere, 267–268
London forces, 130
Luminous intensity, 7; *See also* Light

M

Manometer, 128
Mass: calculating, 9; conservation of, 24; defined, 1; in ideal gas law, 143; measuring, 7; molar, 100–104; in solution concentration, 150–151; stoichiometric

Reaction order, 173
Reaction rates: complex, 175–176; factors affecting, 171–173; laws, 173–175; model, 169–171
Reaction spontaneity, 166–168, 217
Redox reaction: balancing equations, 205–209; half-reactions, 210–212; oxidation and reduction, 201–204
Reducing agent, 201
Reduction, 201
Reduction potential, 214
Representative elements, 54, 63
Research, scientific, 4
Resonance, 85–86
Reversible reaction, 177
Ribonucleic acid (RNA), 249–250
Roentgen, William, 253
Rounding, of numbers, 17–18
Rowland, F. Sherwood, 3, 4
Rutherford, Ernest, 32

S
Safety, laboratory, 4–5
Salinity, 266
Salt, 198
Salt bridge, 213
Saturated hydrocarbon, 226
Saturated solution, 147
s-block elements, 63–65
Schrödinger, Irwin, 44
Scientific law, 4; See also specific
Scientific methods, 2–4
Scientific notation, 11–14
Scientific research, 4
Second, 7
Secondary batteries, 218
Sigma bonds, 80
Significant figure, 16
Single-replacement reaction, 93–94
SI (Système Internationale d'Unités), 7–10

Solids, 22, 131–132, 167
Solubility, 147
Solubility product constant, 184–186
Soluble substance, 147
Solute, 95
Solvent, 95
Solutions, 26; behavior of, 147–150; colligative properties of, 155–157; and entropy changes, 167; heterogeneous mixtures, 157–158; molality of, 154; molarity of, 152–154; and mole fractions, 155; percent by mass of, 150–151
Solvation, 147
Specific heat, 159
Spectator ion, 95
Standard enthalpy (heat) of formation, 165
Standard hydrogen electrode, 214
States of matter, 22; and entropy changes, 166; forces of attraction, 130; liquids, 22, 131–132; phase changes, 133–135; solids, 22, 131–132, 167; See also Gases; Matter
Stereoisomers, 229
Steroids, 248
Stoichiometry, 113–115; calculations, 115–119; gases, 144–146; limiting reactants, 120–122; percent yield, 122–124
Stratosphere, 263, 264–265
Strong acid, 190, 196
Strong base, 191, 196
Strong nuclear force, 255
Structural formula, 83–85
Structural isomers, 229
Sublimation, 134
Substance, 21
Substituent group, 222

Solving Problems: A Chemistry Handbook